Dog Is Love

DOG IS LOVE

Why and How Your Dog Loves You

CLIVE D. L. WYNNE

HOUGHTON MIFFLIN HARCOURT
Boston | New York | 2019

hmhco.com

Library of Congress Cataloging-in-Publication Data
Names: Wynne, Clive D. L., author.
Title: Dog is love : why and how your dog loves you / Clive Wynne.
Description: Boston : Houghton Mifflin Harcourt, 2019. | Includes bibliographical references and index.
Identifiers: LCCN 2019002729 (print) | LCCN 2019003934 (ebook) | ISBN9781328543981 (ebook) | ISBN 9781328543967 (hardcover) |
Subjects: LCSH: Dogs. | Dogs — Psychology. | Human-animal relationships.
Classification: LCC SF427 (ebook) | LCC SF427 .W96 2019 (print) | DDC 636.7 — dc23 LC record available at https://lccn.loc.gov/2019002729

Book design by Kelly Dubeau Smydra

Printed in the United States of America
DOC 10 9 8 7 6 5 4 3 2 1

For Sam
Who makes his dog — and his father — proud

CONTENTS

Dog Is Love

INTRODUCTION

RECENTLY I TOOK some time away from my adopted country, the United States, to visit my native England. It was wintertime, late afternoon, and the sun had already finished its short duty for the day. Along with thousands of others returning home from their day's work in the city, I was coming down the steps at a train station in the outer suburbs of London. These Victorian stations must have looked grand when they were built, and some of them still do in summer light, but at the end of a cold, dank day like this one, they are distinctly depressing: the old, dark red bricks illuminated only by dim and flickering fluorescent lights, the whole triumphant setting infused with the miserable mood of the weary commuter.

As if the scene were not dismal enough, suddenly the station rang out with the urgent barks of a dog. Down at the bottom of the steps, just behind the barriers that prevent people from getting on the trains without a ticket, a young woman — a child, really — was holding on with all her might to one end of a leash. Its other end held a small but noisy and highly energetic dog, most likely a terrier of some kind. This little dog was yapping up a considerable storm.

My immediate unconscious reaction was irritation: an annoying soundtrack had been added to an already gloomy scene. But as I got closer and saw how happy this dog was, an involuntary smile crept across my face.

The dog had recognized somebody in the great human crowd. As that person got closer, the dog's barking morphed from an angry snapping into a sort of happy, almost-howling cry. Her claws skidded over the smooth floor as she struggled to get to her human. When the man was through the ticket barrier, the dog jumped up into his arms and kissed his face. I was only a little way behind and heard the man cooing to the dog to calm her down: "It's OK, it's OK — I'm back now."

Looking around, I saw the whole sea of human faces mirroring my own emotional reaction. First irritation — another tedious burden added to the tired tail end of the day — then involuntary happiness at the dog's love for his master. Smiles spread across the crowd; here and there, gentle laughter followed. People who were traveling with companions exchanged nudges and a few words. The majority of solo travelers tucked their smiles back into their pockets, but a light spring in their steps remained as a reminder of the unexpected small pleasure they had experienced at the station on their way home.

As I took in this happy scene, I was transported in memory to one of my first trips back to the UK after I had first left its shores over thirty years ago. Back then, our family dog, Benji, was still alive. My mother drove to the train station on the Isle of Wight, where I grew up, to collect me, with Benji sitting up alert on the front passenger seat. Since people in the UK drive on the left side of the street, in British cars the positions of driver and front passenger are reversed, compared to those in the United States. This meant that, to my tired and jetlagged eyes, accustomed to seeing drivers where now Benji sat, my dog appeared to be driving the car. My confusion barely had time to register when the car pulled up at the curb, and I opened the passenger door to meet Benji's paroxysm of joy at seeing me again. As soon as he saw me, Benji went crazy with pleasure, just like the terrier at the train station many years later—and just like me, although I kept my emotions under tighter control.

At first glance, Benji may not have seemed particularly special; he was just a fairly small black-and-tan shelter mutt. But he was very special to us. Dabs of sandy-brown color around his eyebrows made his eyes especially expressive—particularly when he was puzzled. We loved teasing him, and he seemed to take all our pranking in good spirit. He could prick up his ears to show curiosity. With his tail he could express happiness and confidence, and he showed his affection with licks from his tongue (which felt like wet sandpaper and elicited protests from my brothers and me, although we felt honored by his attention).

My childhood dog, Benji, sometime in the early 1980s

Benji, my brothers, and I all grew up together in the 1970s on the Isle of Wight, off the south coast of England. When my younger brother and I came home from school, we usually would plunk ourselves down on the sofa, whereupon we would hear and then see Benji racing in from the back garden. Ten feet away, he would launch himself into the air and land directly on top of us, thwacking us with his tail and kissing us each in turn, his little body practically in spasms of joy at being reunited. He clearly loved us—or at least, that seemed indisputable to us at the time.

Many years passed. Benji's short life ended; I busied myself with my wandering life. But my memory of the dog of my childhood endured, as did my fascination with the minds of species other than our own.

In time, I gravitated toward academia, where I came to study how different kinds of animals acquire knowledge and how they reason about the world around them. I wanted to understand how animal minds differ from our human minds. To what degree are human abilities to reason, think, and communicate special to us, and to what extent are they shared by other species on this planet? People are often interested to know whether there are thinking beings on other planets, but it was the other minds on *our* planet that I wanted to learn about.

As a professor of animal psychology, my research at first focused on the most common residents of any lab in this field: rats and pigeons. And for a decade, I lived and worked in Australia, where I was able to do studies on really cool marsupial species that no one had looked into before. It was a great life, full of fascinating intellectual puzzles and interesting discoveries—yet I wasn't completely satisfied.

In time, I realized I wasn't interested in animal behavior in isolation. Rather, I was drawn to the relationship between people and animals. And, of all the thousands of animal species on this planet, none shares a stronger and more interesting bond with our own than do our dogs.

In retrospect, I'm embarrassed that it took me so long to realize that I needed to be studying dogs. Their behavior is so rich: there are dogs that sniff out cancer and contraband, dogs that console trauma survivors, and dogs that help blind people cross busy city streets.

And dogs and humans go way back. Indeed, there is no animal with whom people have had a longer or deeper relationship.

People and dogs have been living side by side for more than fifteen thousand years. This long shared history has intertwined dogs' minds with our own in ways we are only beginning to understand. Partly this lack of understanding is due to simple neglect; when I started studying canine behavior, scientists were only just beginning to show renewed interest in dogs after having ignored them for half a century. This resurgence of attention generated some fascinating findings about dogs — research that would soon set me off on a scientific quest of my own.

In the late 1990s, the field of canine science was gripped by new research that claimed to show that dogs have a unique form of intelligence. Scientists theorized that, over the thousands of years that dogs have lived in close proximity to humans, they evolved unique ways of understanding people's intentions, allowing for rich and subtle communication between our two species. This so-called genius of dogs was heralded as the special quality that made dogs such perfectly suited companions for people, and was thus thought to be a key to understanding and managing our relationship with them.

This theory — that dogs have cognitive abilities that enable them to understand humans in ways no other animal can — still has many supporters among those who make the behavior and intelligence of dogs their business and passion. When I first heard about it, it seemed to me a plausible explanation for dogs' stunning success on our human-dominated planet. And yet, as my students and I began studying dogs' behavior for ourselves, these much-vaunted, supposedly unique cognitive skills seemed to disappear like a mirage each time we reached for them.

I began to wonder: What if dogs didn't have any unique cognitive abilities but rather distinct abilities of an entirely different sort? What kind of talent might that be? And if dogs were special for some reason other than their intelligence, what implications would that hold for the way we interact with dogs and how we take care of them?

These questions didn't come to me all at once. Like most working scientists, I was preoccupied with the research in front of my nose. Sometimes professional expertise may make it more difficult to spot what a layperson might identify straightaway. So at first, I failed to see that, for as long as I had known them, dogs had actually been quite frank with me about their true nature. Benji, my childhood dog, along with the blissfully yapping terrier in the dismal train station a few years ago: with every wag of their tail and lick of their tongue, they had been answering the question of what makes dogs special. The real question was, could a scientist see it?

The study of dogs has undergone a revolution of sorts over the past ten years. Researchers are rediscovering a rich tradition of canine science and reapplying to it the time-tested tools of psychology, as well as the latest methods and technologies from neuroscience, genetics, and other cutting-edge scientific fields. The result has been an explosion of evidence for how dogs think and feel — data that has in turn allowed scientists like me to consider questions that a few years earlier we might never have dared to contemplate, much less commit to studying for years of our professional lives.

My research and the work of many others in the exploding field of canine science make abundantly clear that, although dogs' intelligence does not set them apart from other animals, there is nonetheless something remarkable about our canine friends. This research is perhaps no less controversial and astounding than earlier studies on canine intelligence, because it points to a simple but mysterious source for dogs' unique bond with humans. This phenomenon is perplexing and can cause a scientist to feel conflicted — but it is immediately recognizable, perhaps even self-evident, to any dog lover.

Dogs have an exaggerated, ebullient, perhaps even excessive capacity to form affectionate relationships with members of other species. This capacity is so great that, if we saw it in one of our own kind, we would consider it quite strange — pathological, even. In my scientific writing, where I am obliged to use technical language, I call this abnormal behavior hypersociability. But as a dog lover who cares deeply about animals and their welfare, I see absolutely no reason we shouldn't just call it love.

Many dog lovers casually throw around the L-word, and in my home life I have long done the same. But as a scientist, it has not been nearly so easy for me to deploy it. That is because the very notion that animals have emotions has long been anathema to most people in my line of work. The concept of love in particular seems too soppy and imprecise for the hard-nosed business we're in. Attributing it to dogs also risks anthropomorphizing them — that is, treating them as humans rather than as a species unto themselves. This is something that scientists, rightly, have long resisted, both for scientific accuracy and also for animals' welfare.

Yet I have become convinced that, in this regard at least, a hint of anthropomorphism is permissible, even proper. Acknowledging dogs' loving nature is the only way to make sense of them. What's more, ignoring their *need* for love — yes, as I'll explain shortly, dogs do need love — is as unethical as denying them a healthy diet and exercise.

I have been pushed toward this conclusion by a range of evidence coming out of labs and animal sanctuaries around the world, evidence that very clearly shows that dogs feel love just as we humans do. And once I started looking, I realized that dogs' passion for people shows itself in many ways. We've all heard stories of the amazing feats that dogs undertake to protect their owners. Research into how dogs respond to people in distress makes clear that they do show concern for their humans, even if their actual abilities to offer aid are not nearly as dramatic as Hollywood would have you believe. Even more impressive are studies that show how dogs' and their owners' hearts beat in synchrony when they are together, mimicking the synchrony we find in loving human couples. When they are with their special humans, dogs also experience neurological changes—including spikes in brain chemicals such as oxytocin—that mirror changes we humans experience when we feel love. Indeed, dogs' powerful love for people can be traced to the most minute level of their being: their genetic code, which today is divulging incredible revelations about this species' mind and evolutionary history, which scientists are rushing to process.

These and other exciting recent discoveries have forced me to realize that love is the key to understanding dogs. I also have come to believe—and in the pages ahead I will share ample scientific evidence to support this belief—that it is dogs' desire to form warm emotional bonds, and not any kind of special smarts, that has made their species so successful in human society. Their loving nature makes them so engaging that many of us can't help but return the favor and provide solace to the mutt who shows up on our doorstep, the purebred we bought from a breeder, or the dog at the local shelter who pleaded to be taken home.

Truly, dogs' love is the cornerstone of the dog-human relationship, whether we choose to recognize its significance or not. And I argue that we have a responsibility to recognize it—and also to modify our behavior in light of the evidence for dogs' capacity for love. For the theory of dogs' love (a term I use only half-jokingly) holds the key not only to better understanding these amazing animals, but also to managing our relationship with them more successfully. If dogs' ability to love is what makes them unique, it also stands to reason that it gives them unique needs. And if there is a single simple conclusion to be drawn from my research, it is that we humans need to be doing much more to honor and return our dogs' affection. Their ability to love us simply demands reciprocation—and many humans willingly oblige, even if they have no idea of the science behind this age-old dynamic of mutual adoration. Science can both explain our close relationship with dogs and make it better. We can boost our dogs' well-being with interventions as simple as touching them more, leaving them alone less, and giving them the opportunities they need to live in a network of strong, emotionally positive relationships.

We are living in exciting times in the science of dogs. Genetics and genomics, brain science, and hormonal research are all racing ahead to shine light on questions that many scientists haven't even asked yet: How are our canine companions capable of building such exceptional bridges of affection between species? What conditions are needed in a dog's life to ensure that bonds of affection are forged? How did the dog develop this ability in such a relatively short (by evolutionary standards) period of time? Answering these questions has been the goal of some of the most exciting studies conducted in recent years by pioneering scientists on the frontiers of modern canine research. In this book, I will describe their findings alongside my own.

But it isn't enough just to study dogs and understand them. We need to take that knowledge and help dogs lead richer, more satisfying lives. Dogs trust us, yet in so many ways we let them down. If this book has any value at all, it will be to bring people to the realization that our dogs deserve better. They are entitled to more than the isolated, unhappy lives to which we too often consign them. They deserve our love, in return for the love they give so freely.

These are not just my deeply held beliefs as a dog lover; they are also my reasoned conclusions as a scientist, with data to back them up. As someone who was once himself guilty of dismissing the idea of dogs' love as abject sentimentalism, let me reiterate that, after many years and against my own inclinations, I have found a tremendous amount of evidence to support the theory of dogs' love, and very little that undermines it. That's not soppiness — it's science.

I sometimes feel slightly self-conscious that, after so many years of studying animal intelligence in a ruthlessly skeptical way, I have ended up advocating a view of dogs that some people nevertheless might consider saccharine. But I can live with that because I firmly believe that dogs will be better off if only more people can be persuaded to adopt it.

It is also tremendously satisfying for me to know that what I experienced with Benji all those years ago was the real thing. Love was the true essence of that relationship, as it is of nearly every interchange between dog and human. A lot of dog lovers have known all along that researchers were barking up the wrong tree when they insisted that dogs' specialness lies in their smarts, not their hearts. Science, at last, is catching up.

1

XEPHOS

THE FIRST TIME I saw Xephos, she seemed so tiny. Partly, this was her doing: her little frame was curled up into a frightened ball on the concrete floor of her pen in the humane society's animal shelter. All around her, other, bigger dogs were bouncing up and down in their kennels, shouting to get my attention. But poor Xeph' had hunkered down and was too scared to do more than peek out from behind her hind leg at her unfamiliar visitor.

The shelter was clean, and the volunteer who ushered me through to the kennels radiated concern for his canine charges — but still, it was hard not to be depressed. Xephos's home was a bare, prison-like world of metal bars and stark hard surfaces: a noisy, featureless expanse of concrete and steel. The racket from her neighbors was exhausting. I just wanted to get out of there, and I'm sure Xephos and the other dogs did too.

I had come to this animal shelter in North Florida with my wife, Ros, and my son, Sam, because they had decided to "surprise" me with a dog for my birthday. I use the quotation marks because, wisely, they let me in on the secret. No one should ever truly surprise a loved one with the gift of a live animal; the responsibility of caring for another living being is far too great. After I agreed with their idea, Ros and Sam did, however, take on all the responsibility of finding me a suitable pooch so that I would have the sense of receiving a gift.

In 2012, when we finally decided to adopt a dog, I had been studying dogs in a scientific capacity for several years without actually having one of my own to come home to. With major international moves and parenthood, my life had seemed too complicated to add canine companionship to the mix. Much as I had appreciated sharing my home with a dog in the past, I didn't think it would be right to subject a dog to our unpredictable schedules and frequent absences. I didn't, and still don't, believe that every human life contains a dog-shaped space for a pup to slip into.

But eventually it had become clear that my family could readily accommodate a dog. What's more, I had really begun to yearn for one. Spending so much time during my working days around people and their dogs, or at the animal shelter seeing all the great dogs that needed homes, I felt strange coming home to a dogless domicile. Sensing my yearning, and also pining for a dog themselves, Ros and Sam had taken it upon themselves to find one.

Since they were trying to maintain the element of surprise, Ros and Sam avoided asking for my help, and therefore they ended up looking for a dog at a shelter that I wasn't very familiar with. As a canine scientist who specializes in the study of dogs' behavior, I had conducted research at many different shelters in this part of Florida. But my colleagues and I had skipped over this particular humane society because many of its residents had serious behavioral problems that we had deemed too risky for the young students who helped conduct our experiments. Any dogs that had entered this shelter with an understanding of how to communicate their gentle intentions toward people had all long since found homes. Thus the shelter—a no-kill establishment—was largely left with a population of canines that did not know how to behave in the way that humans desired. Whether or not they were truly dangerous, these poor animals clearly had no idea of how to express to people that they would make good companions.

This sad situation announced itself even before you stepped inside the shelter. The main kennel block was so loud, you could hear the barking cacophony from the parking lot. Once you met the dogs themselves, they exhibited behaviors that seemed the very opposite of welcoming. My colleagues and I had the utmost respect for this shelter's mission and its refusal to euthanize any animals that passed through its doors. But we nevertheless did not feel that we could carry out research there, solely out of concern for the safety of our students. Thus I would not have thought to go looking for a dog there if I had been in charge of the search—which, happily, I was not.

The day before our visit, Ros and Sam had made a reconnaissance trip to this shelter—and they had been eager to come back, for one simple reason. As luck would have it, just the day before my wife and son visited, the shelter had taken in a new pup. This dog was still in the somewhat quieter (if still plenty loud) quarantine section of the shelter and hadn't yet been put into the main kennel block.

Ros and Sam came home very excited about the small black dog they had found. The next day, puzzled that they had apparently discovered such a gentle-sounding animal at this shelter I knew only as a warehouse of dogs serving life sentences, I went along with them to meet Xephos.

And she was such a poor timid wee thing. About twelve months old when we found her, she seemed much younger. Unlike the other dogs in the room where she was held, she whined more than she barked when we came in, and once let out of her kennel, she rolled onto her back and peed herself a little in desperation to communicate her deference to us. She kept her tail tucked as tightly between her hind legs as it is possible for a dog to do. She licked our hands, and when we came down to her level, she wanted to lick our mouths. She deployed the whole toolkit of canine behaviors designed to show respect and the desire to form an emotional bond. She seemed to be saying, as powerfully as she knew how, "I'm your dog. Take me home and I'll love you loyally." It was a compelling argument, and we signed her up straightaway.

We later learned that Xephos had had a tough first year of life.

She had been born at another shelter in the city. Her mother had been abandoned, pregnant, and the litter picked up every bug that was going around. In time, Xephos became healthy and found a human home. But her first family had decided not to keep her. So Xephos ended up back at another shelter, alone, scared, and desperate for a second chance.

By this stage I knew enough about dogs in shelters to know that Xephos's story was sadly common, and that the vast majority of dogs end up homeless through no fault of their own. But once we got her home, I still couldn't help but watch and wait to see what inexcusable bad habit might have caused Xephos to be abandoned by her first human family. Nothing of the sort ever materialized. It was the first of many pleasant surprises this exquisite little creature would bring—and of the many lessons that she would teach me.

As of this writing, Xephos is about eight years old. She remains as charming and easy to live with as she appeared to be when we first met her—perhaps even more so. Gradually, over her first few weeks with us, she shed her shyness, and a strong and happy personality took its place. Notwithstanding her deep black color, she brightens whatever room she is in. No longer is she a timid pup, with her tail tightly tucked; nowadays a visitor is very unlikely to see that appendage in anything other than a proud, upright orientation. She's such a larger-than-life character that I'm often surprised by how small she is physically. She is always the first to greet people at the door: barking up a storm when she hears footsteps getting close and the doorbell ringing, then crying with pleasure when the door opens to someone she knows. She also knows the sounds of her best friends' cars, and she whines instead of barking as they walk up to the door.

In everything she does with people, Xephos radiates affection. Even knowing what I know now about the cause of her companionability, I cannot help but marvel at it. But when we brought her home, her affectionate nature did not make nearly as much sense to me—or seem nearly as miraculous—as it does today.

Of course, I had lived with dogs before and knew how warm their response to our species could be. And yet, as a scientist studying dogs' behavior, I had no frame of reference for this apparently emotional aspect of dogs' lives. The idea that dogs were capable of love — or indeed, any emotion — was, around the time we found Xephos, anathema to canine psychologists like me. Indeed, it was so outside the terms of scientific discussion about dogs that it did not even occur to me to think about it.

Yet by this point in my professional life, I had also begun to question other aspects of the received wisdom about dogs' cognitive capacities. Before long, this skepticism would lead me to a crisis of conscience about the inner lives of dogs, and what makes them who they are. This reckoning, in turn, would set me off on a journey of discovery that fundamentally changed my relationship with dogs — not only Xephos, but also those unfortunate canines still locked away in shelters, and the entire amazing species, at once familiar and misunderstood, of which they are a part.

Xephos came into my life at a critical point in my thinking about dogs. I was struggling to reconcile my scientific research about canine cognition with a set of ideas about the reasons for dogs' success in human society, which had become widely accepted by the time Ros, Sam, and I brought Xephos home in 2012. These ideas purportedly explained the underpinnings of relationships like the one we were now embarking on with this furry little member of the family.

In the late 1990s, when it seemed that researchers had almost entirely forgotten the willing subjects resting at their feet, two scientists had reignited interest in the psychology of dogs by independently proposing a new way of understanding this species and its special relationship with humans. Ádám Miklósi at the Eötvös Loránd University in Budapest, Hungary, and Brian Hare, then a student at Emory University in Atlanta, Georgia (now a professor at Duke University in North Carolina), came from completely different backgrounds, but ended up with the same conclusion: that dogs possess a unique form of intelligence that makes it possible for them to get along with people in a way that no other animal can.

At first, Hare had been investigating the social intelligence not of dogs, but of chimpanzees. Since they are our closest surviving relatives in the animal kingdom, chimps are a natural go-to species for anyone looking to understand what makes human cognition unique. Hare was fascinated by the age-old puzzle of what makes humans stand out in the animal kingdom. At least since Darwin, scientists have struggled to figure out just exactly what the distinction is between the human mind and those of other species. A typical approach to this question runs like this: if you think you have found something that only humans can do, test chimpanzees; if chimps cannot do it, it is unlikely that any other species less closely related to humans can do it either.

At the time, Hare was testing for an ability that seems very simple to us humans. If I know, but you don't, where something you want has been hidden, then I can communicate its location to you by pointing at it with my hand. Hare wanted to know whether this is a uniquely human form of social understanding, or whether chimps might also comprehend the implications of a basic pointing gesture.

Hare's experiment was simple. He would take two upturned cups and, using a screen so the chimp couldn't see, hide a piece of food under one of them. Then, after taking the screen away, Hare would point at the cup with the hidden food. If the chimp chose the cup with the food tucked underneath, it would suggest that it understood the meaning of the human's gesture.

As it turned out, Hare's chimpanzees chose pretty much at random. Simple as the task sounds, it was too much for them to figure out.

Hare thought the chimps' failure was odd because he felt sure that his dog at home could easily succeed at this task. But when he said as much to Michael Tomasello, his mentor, Tomasello assured him that there really wasn't any chance that walnut-brained dogs would succeed where chimps had failed.

And so it came to pass that, the next time he was home with his childhood dog, Oreo, Hare found himself standing in his parents' garage with two upturned cups on the floor, one on each side of him. His dog waited patiently while Hare hid a piece of food under one cup and pretended to hide a piece of food under the other. Then he pointed at the cup with the food in it, and, without any hesitation, Oreo trotted straight to the baited cup.

Hare was convinced that his dog was not just sniffing out where the food was hidden. After all, Oreo didn't know where to go when Hare stood between the two cups without pointing at one of them. It really seemed that Oreo was able to understand Hare's pointing gesture — and thus that the small-brained family pet was succeeding where the much-bigger-brained and closer human relative, the chimpanzee, had failed.

From there Hare went out to a wolf sanctuary in Massachusetts and gave a similar test to several hand-raised wolves. All dogs are descended from wolves, so by testing their wild relatives, Hare was checking to see whether dogs' ability to succeed with this task was something they had inherited from their ancestors, or a capacity that had arisen for the first time in the evolution of dogs.

The results of Hare's wolf study suggested that dogs are, indeed, quite special in this regard. He found that the wolves, unlike dogs, had no idea what his pointing gestures meant. When confronted with Hare's pointing gesture, dogs' wild cousins were just as clueless as the chimpanzees.

On the other side of the world, the Hungarian scientist Ádám Miklósi was independently conducting almost the exact same experiment as Brian Hare — and producing much the same findings. Whereas Hare's path to studying dogs might be termed "down from the apes," Miklósi's could be labeled "up from the fishes." Miklósi had trained in Hungary as an ethologist — a scientist focusing on the behavior of animals in their natural habitats — and originally, the lab he worked in studied small fish. But in the mid-1990s, the director decided it was time to investigate an animal of more direct relevance to many people's lives, which is how Miklósi came to be studying dogs instead of fish. His research group was interested in whether dogs and humans had evolved psychologically and behaviorally to be able to understand each other. Not knowing what Hare and Oreo were up to in Atlanta, Miklósi and his students independently went through the exact same process in Budapest. They first tested pet dogs' ability to follow people's pointing gestures, and found them to be highly successful. They then hand-reared some wolf pups in their homes in Budapest and found that the wolves failed to follow their hand movements to find food.

After analyzing these studies and others, Hare concluded that dogs have a genetic predisposition, bred into them over the thousands of years they have lived among people, to understand people's communicative intentions and comprehend something of human social intelligence. This ability, Hare argued, is the birthright of every dog pup, and develops spontaneously in each and every one of them, even without having any experience of people and the things we do. Hare didn't deny that it might be possible, with tedious training, to teach members of another species to emulate aspects of what dogs can do, but in Hare's account, only dogs are *born* understanding people in this way—the crucial difference between them and every other nonhuman animal on the planet.

When Hare first published his conclusions in 2002, I was really excited—and I also was at a point in my career when I was ready to be energized by something new. That year, I had arrived in the United States as a junior professor in the psychology department at the University of Florida. I had spent the preceding decade on the faculty of the University of Western Australia, where I had studied the behavior of marsupials such as the fat-tailed dunnart—a gorgeous little mouselike animal with less than one-tenth of an ounce of brain tissue, but a really quick learner. The move to Florida was exciting, but it meant cutting myself off from the marsupials that had fascinated me. I hadn't yet thought to turn my attention to dogs, but as I read Hare's research, I was fascinated by the idea that a canid with no special endowment in the brain department had somehow acquired forms of cognition that were otherwise known only in our own notoriously cerebral species.

Hare's research started to appear in the scientific literature at around the same time as the first papers offering genetic analyses of dogs' DNA. The geneticists' input added an additional fascinating layer of complexity to the discussions of what made dogs unique.

Geneticists estimate how old a species is by comparing its genetic material to that of closely related species, and studies from Sweden, China, and the United States made clear that the process of domestication that created the dog was, by evolutionary standards, extremely rapid. Rather than the millions of years needed for notable change in relatively large and long-lived species such as the dog's most immediate ancestor, the wolf, dogs had appeared on the scene within, at most, a few tens of thousands of years. Wolves usually breed only once a year and don't reach sexual maturity until their second year of life. That may sound young to us, but compared to most animals it is a very slow life cycle. Speed of evolution is necessarily tied to how long it takes individuals to produce the next generation of their kind, so an animal that can produce a new generation only every two years would be expected to evolve very slowly.

These two parallel threads of research became interwoven in my mind. If dogs were truly blessed with a unique ability to innately understand humans, as Hare claimed, then they must have acquired this power in the blink of an evolutionary eye. How, I began to wonder, might they have gained this ability so quickly?

Just as this question was taking shape in my mind, the perfect student came along to help me answer it. Monique Udell had a background in both psychology and biology and a tremendous capacity for sheer, bloody hard work. Crucially, she also was willing to risk starting work on a PhD with a mentor who wanted to research a species he had never studied before. Working together, Monique and I began to explore the implications of these exciting new findings about canine evolution and cognition.

We started out by repeating Miklósi and Hare's pointing experiment with some pet dogs in their homes. It was pretty easy to do, and the results of our study exactly matched those of Hare and Miklósi: pet dogs really are exquisitely sensitive to human actions and intentions. We would hide food under one of two containers on the floor, and when Monique pointed at the one that hid the treat, the dogs went trotting off to precisely that container. It was as if they had read the scientific papers too.*

Although we found results that matched exactly what Hare and Miklósi had said about dogs, we hadn't answered our bigger questions: What, if anything, drove dogs' rapid evolution of the ability to understand human gestures? How had dogs acquired this skill?

No sooner had Monique and I turned our attention to this problem than an opportunity to investigate it presented itself, in the form of an invitation from the administrators of Wolf Park, a research facility in Indiana. They wanted us to come out and test their wolves.

It was not an excess of physical courage that attracted me to the life of a university professor, so I'm not embarrassed to admit that I experienced a fair degree of trepidation as I sat in the education building at Wolf Park, listening to the head curator, Pat Goodman, delivering the compulsory wolf safety lecture.

The rules for engaging with the denizens of Wolf Park are pretty straightforward. You shouldn't stare directly at a wolf, but neither should you take your eyes off it for a moment. It's important not to

---

* You can easily try this yourself, with your own dog. It will probably work best if you have a friend who can wrangle the dog for you while you bait a container. Some dogs are a little nervous about knocking over upturned plastic cups to find what is underneath them, but the study will work just as well if you don't actually hide any food inside anything. Just put a piece of food on top of the container you pointed to after the dog has made its choice. You will likely find that your dog goes to the location you point at most of the time.

make any sudden moves, but just as important not to stand still with your hands hanging uselessly by your sides. If you are too immobile, the wolves might mistake you for a chew toy, Pat explained, which was not exactly reassuring. But the really important thing, Pat made clear, is not to trip on a log or a rabbit hole. Apparently it's very difficult to pull a wolf off somebody.

Thoroughly rattled by this hour-plus rundown of the really bad things that a two-hundred-pound gray wolf can do to a puny psychology professor, I was finally ready to meet my research subjects. It was time to bundle up against the cold September day and head down to the wolf enclosure.

Wolf Park is an oasis of pleasantly rolling country in the vast flat expanses of central Indiana. Right up to the entry to the park, there is nothing but plains, but the land upon which the park itself sits provides a welcome break in topography, with a creek, some wooded corners, and a nice big lake for the wolves to play in. As one of the few patches of trees amid thousands of acres of soybeans and corn, the park also acts as a refuge for birds, which add a happy soundtrack to the beautiful scene. It really is a gorgeous place—but I have to confess that I'm not sure how much of that I took in on our first visit. Mostly I was focused on the large carnivores whose home I was about to enter.

The moment of truth—and terror—finally arrived when Monique and I entered the wolf enclosure. No sooner had I stepped through the gate in the chainlink fence than one of the older wolves, Renki, bounded over to me. Before I could even get my hands out of my pockets, he plunked both his forepaws on my shoulders.

I just had time to think, "So long, sweet world," before Renki licked me powerfully on each cheek.

In an instant, I knew what acceptance into a wolf pack feels like—and I can tell you, blessed relief is no small part of it. I stood around for a while longer, getting to know my new pack mates and research subjects. Finally, once I felt reasonably comfortable around the wolves and it was clear that they did not resent my presence, I set about conducting the test that had brought me to Wolf Park in the first place.

The author's initiation into the pack at Wolf Park

Monique and I had been invited out to Wolf Park because the staff there had heard about the new research from Brian Hare's and Ádám Miklósi's labs. Specifically, they had taken notice of—and issue with—the claims that dogs have a unique ability to follow the implications of human actions: an ability that, according to Hare, dogs share with no other animals, wolves included.

There can be few people on the planet with a more nuanced understanding of the behavior of wolves than the staff and volunteers at Wolf Park. Since 1974 they have been hand-rearing wolf pups, acting as surrogate parents to them and bringing them up so that these wild animals accept people as social companions. The head curator, Pat Goodman, and Wolf Park's founder, Erich Klinghammer, perfected the techniques, which involve keeping a human "mother" with the pups twenty-four/seven for the first several weeks of life, so that the wolves grow up to see people around them as part of the social fabric of their lives. Pat and many of the other Wolf Park personnel also have dogs at home, so they spend their working hours with wolves and their time off with dogs—an arrangement that gives these staffers a well-informed sense of the similarities and differences between hand-reared wolves and dogs.

It was these uniquely well-informed wolf-and-dog people who had first made contact with me to suggest that Hare and Miklósi were wrong. These Wolf Park workers had the distinct impression that the wolves they spent their days with were every bit as sensitive to the things people do as the dogs they went home to every evening.

Hare and Miklósi had each run studies with wolves to test exactly this question, of course, and they had independently come to the conclusion that wolves are incapable of understanding human gestures. I had no particular reason for cynicism about their findings, especially since they came from independent labs on opposite sides of the Atlantic. But, at the very least, I thought it would be fun to try the wolf experiment myself. The skepticism of the Wolf Park staff had sparked my curiosity too. Was it possible that the wolves in Hare's and Miklósi's studies—raised by hand in a Massachusetts sanctuary and a bunch of apartments in Budapest, respectively—were not representative of the species as a whole?

I had never seen wolves up close before, and I was tremendously impressed by both their daunting power and their apparent intelligence. These wolves were the size of the largest dogs—I immediately thought of massive breeds like Irish wolfhounds. But unlike large dogs, which tend to be rather slow in their reactions, gray wolves are quick. Really quick. If a rabbit pops up in their enclosure, *bam*—they've got it in an instant. They kill like professionals, with calculation and without remorse.

Just as striking as their lethality is their sociability. The wolves' engagement with one another and the people they know well is rich, and moving to observe. Their amber-golden eyes seem to glow with intense presence in the moment. I felt really privileged to be allowed into their lives.

I also recognized that discretion was the better part of scientific valor. After chatting with the staff, sitting through the safety class, and venturing into the enclosure for our introduction to the wolves themselves, Monique and I opted to not press our luck. We exited the enclosure and allowed people more familiar with the animals to run the first round of pointing experiments for us. Instead of handling the baited cups and performing the pointing actions ourselves, we would shout instructions to three Wolf Park staff, who would carry out the tests. We all agreed this would be safer and more likely to reveal the wolves' true capacities. We hoped that in time, once the wolves had become comfortable with us, Monique and I might be able to handle some of this work ourselves — but on that first visit we wanted to improve our chances of success by letting the wolves, often wary of strangers, work with people they knew well.

Some interns helped clear an unused enclosure of debris, and one by one the wolves were brought in to be tested. Pat Goodman and two other staffers took turns performing one of three roles: standing between two containers and pointing at one of them; standing about ten feet away to tempt the wolf back to the start position after each test ended; and simply hanging around to make sure everyone was safe. Monique and I called instructions through the fence and provided refills of small chunks of summer sausage, which our intrepid collaborators used to reward the wolves for correct choices and to coax them back to the start position after each test had run its course.

It took a while to get going, but once everything and everyone were in their places and the study got underway, Monique and I were quickly stunned: the wolf was every bit as good at this task as were the best-performing dogs.

In an instant, our research had immeasurably complicated what had seemed to be a cut-and-dried distinction between the cognitive abilities of dogs and wolves. For a scientist like me, who lives to turn over stones to see what may be hiding underneath, and whose whole world revolves around finding questions in need of answers, moments like this are a rare thrill. By coincidence, it was my birthday when we first went out to Wolf Park, and this finding was by far the most memorable birthday gift I have ever received — apart from Xephos, of course.

Once I got over the initial excitement of this startling result, we conducted the same experiment on several other wolves at the park. We found the same pattern of behavior over and over again. These wolves could follow human pointing gestures just as well as any dog could.

On our way back to Florida, Monique and I mulled over the possible reasons for the discrepancy between our observations and Brian Hare's theory of dogs' innate "genius." We knew that genius — or whatever you want to call dogs' remarkable sensitivity to people — cannot be chalked up solely to our pooches' evolutionary inheritance. To be sure, evolution (and that special case of evolution we call domestication) is an undeniably important factor, but there is another crucial component underlying everything an animal does, one that shares an equally important role in determining whether dogs, or wolves for that matter, can read intention into human gestures: and that is nurture, rather than nature.

Evolution is the outcome of natural selection, the process by which species change because individual organisms are born with different sets of genetic traits that allow some to survive better than others and produce more offspring in the next generation. Over countless generations, some traits are selected and passed down, coloring the complexion of an entire species with its own unique kaleidoscope of characteristics — among them the anatomical and cognitive peculiarities (such as intelligence) that lay the groundwork for that species' typical behaviors.

Domestication is a special case of evolution, one whose mechanisms have been subject to some debate. Darwin, who introduced the world to the concept of evolution, believed that animals became domesticated when people selected for breeding those beasts that were most useful to them; in time, Darwin theorized, this practice would give rise to an entirely new species. He called the process of domestication *artificial* selection—in contrast to *natural* selection, which is the term he coined for what happens when the forces of nature decide who lives and who dies. Today we aren't as confident that the whole story of domestication can be charged to our own species' account: it seems more likely that some large chunk of domestication was actually natural selection. Whether it is due to natural or artificial selection, however, domestication is a form of evolution—a process through which animals change over generations due to the selection of some individuals to survive, thrive, and pass on their genes.

But evolution alone cannot create a friendly companion animal for a human home. Natural and artificial selection can act on the underpinnings for an animal's typical behavior and its intelligence, certainly—but evolution can never fully account for the unique cognitive and behavioral package (what we often think of as the "personality") of an individual dog. This is because, although evolution lays down the blueprint for a living thing, it cannot control how that blueprint will be read. Each individual animal is built from genetic information read out by the particular experiences that the individual undergoes as it develops. Consequently, evolution on its own cannot make a friendly dog.

Just as the legs that give us the capacity to walk are part of our evolutionary inheritance, so too are the structures in our brains that give rise to our personalities. And what's true for us is just as true for our dogs: they inherit brain structures that prepare them to be able to enter into relationships with people. But the fact that my dog has a relationship with me and is sensitive to the actions of people in her life is not just a consequence of the evolution of her species; it is also dependent on having grown up in a world that provided her with opportunities to develop the qualities that define her as an individual.

Experience, in short, is the other factor that shapes dogs' actions and their minds. This is obvious when you think about it: after all, no puppy or kitten or youngster from any other domesticated species is born tame. Tameness has to be learned by each individual in its own lifetime. The sweetest puppy will grow up to be a wild animal if it is not early in life introduced to people. (Back in the 1960s, experiments were carried out that established exactly this. At a lab in Bar Harbor, Maine, John Paul Scott and John L. Fuller reared dog pups without any contact with humans for their first fourteen weeks of life. Then, when they tested the dogs as young adults, they reported that they were, as the researchers put it, "like wild animals" and could not be approached.)

Biologists call our evolutionary history phylogeny, and our personal life history ontogeny. It is a truism of biology and psychology that each of us is the product of our combined phylogeny and ontogeny. None of us would be as handsome, smart, and charming—not to mention modest—as we undoubtedly all are, were it not for an evolutionary history that set the stage for the life experiences that in turn molded our characters into the enviable shape they now have. The same is true for dogs. Each has the personality he or she possesses—a personality that, in lucky canines, makes them exceptionally well suited to human companionship, with all of its attendant rewards—only because of a rich interplay between their genetic endowment and the world they grew up in.

The idea that dogs' behavior and intelligence result from both domestication and experience seemed pretty uncontroversial to Monique and me when we viewed it in the light of these basic scientific principles—but it had become grounds for a dispute of sorts in the nascent field of canine cognition, a debate that Monique and I unwittingly stumbled into. On the one side were scientists like Hare and Miklósi, who argued that dogs' ability to understand humans was due to a unique evolved cognitive capacity—part of the birthright of every dog and not dependent on any particular life experiences. On the other side were scientists like Monique and me, who believed that appropriate life experiences, as well as the right genetic endowment, were key in giving dogs the ability to be companions to people.

For our refusal to accept that idea that dogs, as a direct consequence of the evolutionary process of domestication, were born with the innate ability to recognize the meaning of the things people do, we got to play the role of behaviorist spoilsports. After we published the results of our study at Wolf Park, a journalist called me the "Debbie Downer" of canine cognition research. That stung.

I had to think about where I'd ended up. How had I, someone who cared deeply about animals' minds and who had devoted my life to studying them, developed such a negative reputation as a *doubter* of their cognition? I felt misunderstood, and more than a little hurt that my affinity for dogs had put me in the position of appearing to disparage them.

I could see how, to people who didn't know me, I seemed to be saying there was nothing remarkable about dogs. But I wasn't trying to deny there was something special about them. Quite the opposite, in fact: dogs' unique bond with humans was what had attracted me to them as research subjects in the first place. Like the dog-loving staff members at Wolf Park, I needed to look no further than my own living room—where Xephos was often ensconced companionably on the couch next to me as I read up on the latest scientific papers and articles in the popular press, tracking the growing furor over Monique's and my research—to find inspiration and motivation for my daily work.

Dogs are unique; about that, I had no question. I was just skeptical about the dominant theory of what it was that made them so special. As a scientist, I was willing to wear this "Debbie Downer" label as a badge of pride; I wasn't going to let myself be railroaded into a view of dogs I just could not accept. As a dog lover, however, I was determined to get to the bottom of what makes dogs unique. As I learned more about dogs' cognition and their lives in human society, I was beginning to realize that the debate gripping this field was not just an academic argument. There was a lot at stake — most of all, for dogs themselves.

In addition to testing wolves' and pet dogs' ability to follow human gestures, Monique and I — along with Nicole Dorey, another good friend and collaborator — had tried the exact same test at an animal shelter near our home base in Gainesville, Florida. And the results were not pretty.

Not a single one of the shelter dogs in this experiment understood what was implied by a gesture toward a container on the floor. Each stared blankly at Monique as she stood between the two containers, waiting for it to make a choice; or the dog came up and sat nicely in front of her, apparently asking as cutely as it could to be given the treats it knew she had; or the pooch simply wandered off, looking for something better to do.

At first, we thought that maybe these dogs had been traumatized in their earlier interactions with people, and didn't trust that Monique was doing something nice for them. But while it is certainly true that shelters hold plenty of dogs that have been let down by our species and whose trust in humans has been betrayed, for our study we carefully selected dogs that clearly were thrilled to bits to be in the company of humans: taken out of the kennel, played with, and offered treats far superior to the usual daily rations. The dogs Monique worked with honestly just didn't seem to comprehend what her gestures meant.

The dominant theory about dogs' uniqueness had grim implications for these uncomprehending creatures. If we believe that all dogs are born with an innate capacity to understand people's actions and intentions, as Brian Hare and his colleagues argued, then dogs that seem unable to understand human intentions would have to have some sort of deep cognitive deficit preventing them from fully realizing their evolved potential as dogs. If the ability to understand human gestures is innate, then the failure to understand them must be innate too. This could lead to the conclusion that dogs like those we tested at the animal shelter were simply less suitable as companions for human beings.

The results Monique and Nicole were getting from our local shelter, where not a single dog could follow their gestures, could result in terrible outcomes for a lot of dogs—both at this facility, where at the time euthanasia of unadoptable pets was still part of standard practice, and at similar shelters across the country, and indeed around the world. Today millions of dogs are sacrificed every year because no home can be found for them. Any qualities that might help determine whether a dog stays in a shelter or goes home with an adoptive family could literally be the difference between life and death. For canine scientists and dog lovers like Monique, Nicole, and me, nothing could be more important than understanding how dogs find fulfilling lives in human homes.

We were determined to understand what was wrong with these poor pups at the shelter, as well as the implications of their handicap. Were they lacking the genes to understand people—that is, had something in their phylogeny made them incapable of interpreting our gestures? Or was the issue instead in their ontogeny, something in their personal history that made them unable to understand Monique's pointing? That would give us an explanation for their deficits. It would also, we hoped, point to a way of correcting the problem.

If these dogs were capable of learning the meaning behind human gestures, we knew a simple dog-training principle that would enable us to teach it to them. Each time you point at something your dog is interested in—a morsel of food, or a ball, or anything else—in order to help her find that valued item, and she locates it successfully, that success will be rewarding. In the scientific jargon, we say the action the dog just performed has been *reinforced*. And everything we know about animal behavior says that reinforced behaviors are more likely to be repeated in the future.

This simple behavioral mechanism, we surmised, may be enough to enable dogs to learn to follow human pointing gestures. If Monique were to point at a treat and the shelter dog she was studying were to find it — even if just by accident at first — that dog might be more inclined to follow her gestures in the future. And if that happened, then it might mean that there was nothing inherently wrong with the dogs at the shelter. Maybe they had been unable to follow human gestures simply because they didn't have much experience with people pointing at things. Maybe they had not had the chance to learn, or had forgotten, what human gestures mean.

All we had to do was go back to the shelter and see whether it was possible to train the dogs there to follow human pointing gestures. We would just need to point at the container with food in it and allow the dog to see what the outcome could be for picking it. If the training didn't work, that would suggest that Hare was correct in claiming that dogs have an evolved, innate ability to follow human gestures — an inherited quality that somehow certain dogs had missed out on. But if it did work, that would suggest that dogs learn to follow human pointing because they have personal experiences of pointing gestures that indicate where reinforcing outcomes are located. In other words, it would imply that dogs' ability to understand human gestures is acquired, rather than inborn — and thus that they are no different than other animals in this regard. The source of their exceptional bond with humans would have to lie elsewhere.

I suggested to Monique and Nicole that they might try working with each dog at the shelter for a whole day to see whether they could teach it what it means when a person points at something. But Monique and Nicole felt that half an hour with each dog ought to be enough, and their intuition turned out to be correct: twelve of the fourteen dogs they tested learned to follow human pointing in less than thirty minutes. In fact, the average time it took the twelve successful dogs to learn to go where someone pointed was just ten minutes. In ten minutes, a dog that had previously had no idea what an outstretched human arm meant was transformed into one who dutifully followed a human gesture.

This was such a thrilling result: these dogs obviously were not beyond saving! This finding also indicated that we needed to double down on our efforts to understand the behavior and cognition of dogs. We clearly had much to learn about what makes dogs such remarkable companions for people. And we also had much to contribute to dogs' welfare if only we could figure out what, exactly, sets dogs apart.

Pointing, of course, is only one of the many ways that humans communicate with dogs. And the kind of social-cognitive intelligence that Brian Hare, Ádám Miklósi, and their colleagues singled out as unique to dogs is only one aspect of what makes people view dogs as special. While Monique, Nicole, and I had shown that dogs' ability to understand human gestures is learned, not inherent, it was still possible that other forms of canine intelligence might help explain the unique bond between dogs and humans in a way that recognition of gestures could not. So before we went any further, we also needed to rule out these other types of intelligence as being unique to dogs.

Ask any dog lover, and she will be able to name at least one dog she has known who was marked by exceptional intelligence. In my case, that particular specimen wouldn't be Xephos (sorry, sweetie!), but rather Benji, the dog of my childhood in England, in the 1970s.

Benji was what many people call a clever dog. Mainly this means that he had a demonstrable ability to escape from the house and back garden and represent his own interests in the world outside. Benji and I went through adolescence around the same time, but whereas I became a spotty, tongue-tied nerd, he was a natural around the ladies. (His collar tag stated, "Hi, I'm Benji. My phone number is Shanklin 2371"—but we used to joke that, if he had anything to do with it, the other side of the tag would read, "Hello, Darling, wot's your name and phone number?" We always imagined him saying this in a cockney accent because we thought of him as a lovable larrikin, a beloved but disreputable character.) Benji was one of those fairly small and very elastic dogs that could squeeze through the smallest gap in a hedge but also jump over some surprisingly high walls. Surely another important factor in his inclination toward extracurricular outings was that we never had him neutered. My mother didn't like the sound of it, and my father never considered anything to do with the dog as his business. So whenever Benji scented a receptive female in the neighborhood, he would hop off looking for trouble and come home a few hours later, looking tired but happy.

Benji's little outings to visit his girlfriends probably correspond most closely to what a biologist would view as intelligent behavior. To a biologist, the urge to reproduce is the one essential drive in life, and any tricks that an individual figures out to help him in that endeavor are what count. But the urge to procreate isn't what most laypeople would think of when they hear the word "intelligent."

Animals in general and dogs in particular obviously possess many other kinds of smarts, ones that are closer to a standard dictionary definition of "intelligence" than the basic urge to go out and mate. Among my personal favorites are sniffer dogs, whose ability to detect things we humans just cannot perceive can seem nearly magical. I am totally in awe of dogs that can detect cancer or improvised explosive devices, for example, just by sniffing the air. If I'm not giving sniffer dogs my personal top spot as the smartest dogs out there, it's only because such a large part of what impresses me about them is really due to their perceptual abilities—their ability to smell things we cannot detect—rather than actual learning skills or intelligence.

The cleverest dog I have ever met, and the one with the most remarkable ability to understand human intentions, has to be Chaser. This isn't just my personal assessment. Nicknamed "the world's smartest dog" by the BBC, this classic black-and-white border collie knows the names of over twelve hundred toys. Chaser is from true working-border-collie stock: a dog that needs something to keep her occupied or she is going to tear up the furniture. Her owner, John Pilley, was a former psychology professor who found himself looking for a hobby in retirement. John had read research from Germany about a border collie who knew the names of more than three hundred distinct objects, and after adopting Chaser — so named for her love of chasing things, naturally — he decided to test for himself the limits of canine comprehension of human language.

When I visited him and Chaser in 2009 at their home in the beautiful Upcountry of South Carolina, John and Chaser had been working together for over three years. John kept an immense store of toys in large plastic storage containers on the back deck of his home. He invited me to go out there and pick ten toys at random. They were the kind of playthings one gives to dogs and small children, and on each toy John, using a permanent marker, had written a name. He asked me to write the names on a notepad, bring the toys into the house, and put them down in the space between the sofa and the back wall of the family room. I did all this while John and Chaser waited outside on the front deck so they couldn't see which toys I had selected.

When I was ready, I called them back inside. John sat on the sofa facing away from the space where I'd placed the toys. He put a large empty plastic container on the floor in front of where he sat and instructed Chaser to sit next to it. Everything ready, John read the first item off the list: "OK, Chaser, go get Goldfish." Chaser looked around, not knowing where I had put the toys. "Goldfish. Go on, Chaser. Get Goldfish."

Thus prompted, Chaser began to circle around, looking for the toys. She quickly found the pile of things behind the sofa and, bringing her snout close to the ground, started searching among them for Goldfish. Aside from seeming slightly shortsighted — she brought her face very close to each object before deciding whether it was or was not Goldfish — she appeared to be doing what any human would do in this situation. Quickly she picked up one of the toys with her mouth and ran with it back to John.

"Put it in the tub," John instructed, gesturing to the plastic container in front of him. This seemed to be the tricky part. Chaser hesitated, apparently reluctant to let go of her find. "In the tub," John repeated. "Put it in the tub." Finally Chaser acquiesced and released the toy into the plastic container.

"OK, let's see," John said, as he lifted the toy out and read its label. He nearly exploded with glee as he confirmed her correct selection: "See. That's Goldfish! It's gold; it's a fish; it's Goldfish!"

And with that, John threw Goldfish across the room and Chaser bounded in delight after it. She brought it back. He threw it again. She brought it back. He threw it again. It was hard to say who enjoyed this little dance more, John or Chaser, but after a few runs back and forth, John instructed her once again to "put it in the tub," and gave her a loving rub round the scruff of her neck before proceeding to the next item.

And so the two of them worked through my list. From Goldfish to Radar, to Wise-owl, to Bling, to Feozies, to Shirley, to Treasure-box, to Chipmunk, to Sweet potato, and finally to Mickey-mouse. In most cases, John rewarded Chaser with the opportunity to chase after the toy once she had "put it in the tub," but sometimes he mixed things up a little by playing tug with the toy. Every time she brought him the correct item, he exploded with delight, and every time he ended the game with a loving tousle of her head and a scruffing of her neck. Seldom has science struck me as more affectionate and enjoyable than watching those two work and play together.

Since Chaser and John were having such fun, I went out onto the back deck and got ten more toys, and we repeated the procedure. Chaser got them all correct, so we did it again. And again. I forget how many times we repeated the name game, but I'm sure I saw Chaser pick up at least a hundred items by name alone. Only once did she make a mistake — or seem to. On closer inspection, it turned out that John had misread my diabolical handwriting and Chaser, not finding what he asked for but not wanting to disappoint, had brought him a different object.

John stopped training Chaser to recognize names for new objects when he got to around twelve hundred items simply because he found he was unable to remember which toys he had already acquired, and was bringing home duplicates. He would happily devise new names for these redundant toys, teach the names to Chaser (she got so good at this that she could learn a new name in a single try), and only by chance would he later notice that he now had two identical objects with two different names. Up until the twelve-hundredth toy, Chaser never slowed down in her learning of new object names.

I encouraged John to publish his findings in the scientific journal of which I was the editor at the time; his report became one of the most widely read papers *Behavioural Processes* ever published. John went on to write a best-selling book that immortalized his wonderful dog, and they even appeared on national TV together before John finally succumbed to leukemia, a few weeks short of his ninetieth birthday, in June 2018.

Chaser's story is only a single data point, of course—but her amazing success in learning so many words, and the fact that she is the only dog John tried this with, suggests that the ability to understand language is latent in any border collie. This is supported by the fact that the dogs in Germany that have vocabularies of dozens or a few hundred item names—among them the dog that was the subject of the research that inspired John's long-running experiment with Chaser in the first place—have also all been border collies.

On the face of it, this certainly seems like evidence that Chaser's breed is blessed with exceptional inherited intelligence. But border collies are also exceptional for another quality: their extraordinary motivation to work. John trained Chaser for around three hours per day for three years to get her to the point where she was so wonderfully fluent in understanding human language. And at least part of the secret of Chaser's success lies in the fact that she found the opportunity to chase things tremendously rewarding; she was strongly motivated to work on learning language with John because the act of finding each toy was inherently reinforcing. Most dogs can be rewarded with food, but there is a limit to how many treats it is wise to put into a dog. Food-rewarded dogs cannot be trained continuously for several hours per day; they would not only get full, but also quickly become overweight. A dog like Chaser, however, who is motivated to work just for the opportunity to chase after a moving object, can be trained for far longer each day. People who work with border collies know that this means they have to be extra cautious about the welfare of the dogs; the animals will literally work themselves into the ground, overlooking injuries, if you aren't careful. It is this boundless energy, this fanaticism, really, that makes border collies the ideal subjects for this kind of project. Few other breeds of dog have that much enthusiasm for work.

What's more, Chaser's skills — while certainly impressive — also were rather simple, and arguably owed more to John Pilley's virtuosic training than to any canine intellect of her own. With time, John's training of Chaser became so smooth and easy that it could seem as if he was just explaining to her what a new object's name was, much as a parent might tell a child the name of something unfamiliar. But the principle in play here is different in an important way.

Here's the scenario: John has a novel object. John can make that object tremendously rewarding to Chaser, either by throwing it (so she has the opportunity to run after it and bring it back to him) or by playing tug of war with her, using the object as their tug toy. The dog loves this game almost as much as chasing. John says something like "Hey, Chaser, go get the Thingamajig" — Thingamajig is the name he is going to teach her for this new object — and with that he throws the Thingamajig as far as he can. Thrilled at this very rewarding opportunity to run after something and bring it back to her master, Chaser bounds off to get Thingamajig and brings it back to John. Then John says, "Give Thingamajig to Pop-pop" (the name he uses for himself when talking to her). Chaser enters that delightful state of ambivalence that dogs who like chasing toys often display when told to give up the precious chased thing. Should Chaser hand it over and get the magical reward of another opportunity to chase after it, or should she hold on to it because it is her prize and she wants it? (There is some risk to the first option, as Chaser knows from past experience: Pop-pop might put the toy away, and the chasing game will be over for a while.) So John keeps cajoling Chaser — "Give the Thingamajig to Pop-pop" — over and over until Chaser hands it over. And then he throws it again: "Go on, Chaser, go get the Thingamajig." And the cycle repeats itself.

In a situation like this, with just one object in play, most dogs would not pay much attention to the unique vocal label the human is using to refer to this object. But by the time I met them, Chaser had over three years' experience of playing this game with John, who added complexity to their play by offering Chaser multiple named objects to retrieve. He rewarded her with the chasing opportunity only if she brought back the right one: the one he named. Over the surely millions of times Chaser played fetch with John, the critical nature of the single new word in unlocking this much-valued opportunity to chase something and bring it back had been drilled into this very attentive dog.

Any readers with a highly chase-prone dog and a lot of spare time on their hands can emulate this pattern of training and see how far their dog's vocabulary can be extended. Unfortunately, my own dog, Xephos, has no interest in chasing toys unless someone chases her to get it back — and I'm just not enough of a fitness fanatic to try to teach a dog a decent vocabulary by chasing her around the backyard for three hours a day.

What, then, does Chaser's training really demonstrate? It shows that she can associate a sound like the word "thingamajig" with an object, and she knows she will be rewarded for bringing that object to John. This form of associating things is understood to be one of the most basic building blocks of intelligent behavior and is seen in all the animal species that have ever been tested for it. It is, indeed, the Pavlovian conditioning that the great Russian scientist Ivan Petrovich Pavlov discovered over 120 years ago — using dogs as his subjects.

What makes Chaser exceptional is the sheer number of different sounds that she can associate with distinct objects. Adding more objects to her vocabulary demonstrates the capacity of her long-term memory, but it does not add any real intellectual complexity to what she is doing. Her immense vocabulary is entirely a testimony to John's patience in training her, and her willingness to keep working hour after hour, day after day, year after year.

This is not to downplay Chaser's achievement—just to put it in context. A vast range of animal species has been shown to form associations, and some animals have been found to perform cognitive feats that are far more remarkable than just associating specific sounds with particular objects (let alone associating a pointing gesture with a morsel of food). Pigeons can identify whether a picture depicts a chair, a flower, a car, or a person; dolphins have demonstrated that they understand grammar; honeybees spontaneously communicate to their hive-mates the distance, direction, and quality of the food source they have discovered on their foraging trips. To my knowledge, dogs have not achieved any of these things.

What's more, many other animals can be trained to form associations between human actions and outcomes, allowing them to appear to be able to "read" the intentions in a human's behavior. Perhaps the most amazing example of this—and certainly my personal favorite—comes from bats. My student Nathan Hall (now a professor at Texas Tech University) carried out a study replicating the work that Monique Udell and I had done showing how dogs follow human pointing gestures, but he worked with bats living at a conservancy in Florida. The procedures were in principle the same ones we used with dogs and wolves—the major difference being that instead of walking on the ground, the bats pulled themselves across the chicken wire that formed the ceiling of their enclosures. Consequently, instead of pointing downward toward containers on the ground, Nathan pointed up at containers hooked onto the wire ceiling.

This experiment was particularly useful for understanding whether it was genetic inheritance (phylogeny) or life experience (ontogeny) that made it possible for an animal to follow human gestures, because around half the bats had been born at the conservancy and reared by bat mothers, while the other half had been dumped at the conservancy by people who had raised them themselves, in the hope of having an unconventional pet. (Like most undomesticated species, bats make rotten pets, and sooner or later their owners had wearied of cleaning up bat poop and abandoned them.) In a finding that offered powerful support to the theory that Monique and I had developed, Nathan found that the bats reared by their own mothers did not follow human pointing gestures, but bats that were reared by humans — and therefore had come to recognize that the motion of human limbs had important implications for them — did follow human gestures.

In the course of analyzing the experiments conducted by scientists like Nathan and John, as well as continuing to conduct our own, my collaborators and I gradually came to realize that what Hare calls the "genius of dogs" is actually present in any animal that has been reared with people from an early age. The ability to follow human intentions cannot therefore be due to genetic changes incurred in the process of domestication; we have since observed this ability in wolves and many other animals that are not domesticated. Rather, we are now convinced that this capacity can develop in any animals that are reared alongside people and are dependent on people for their daily needs.

To be fair, dogs' abilities to detect associations between things we do and consequences that matter to them oftentimes are so subtle that it can seem as though our dogs can read our minds. An older gentleman once came up to me after a talk I gave to a community group: "I thought you might be interested to know," he said, "my dog's psychic." Of course I was curious, though also a little wary. It turned out that the reason this man's dog struck him as having supernatural powers was because his little Westie could always tell whether he was planning to take her for a walk when he got up from his chair, even before he put his shoes on or reached for her leash. Now it's true I never got the opportunity to test this dog, so a small chance remains that his Westie really possessed psychic abilities, but I think it is far more likely that this dog, like my own Xephos, had come to notice the different ways he moves his body when he gets up from a chair to do different things. Xephos seems to know when I rise from my desk chair at home whether I am going to make myself a coffee or take her for a walk round the block. Although I don't notice as I do it, I'm sure that the way I carry myself, and probably even whether I look at her, conveys to her what my intentions are.

Sniffer dogs, with their uncanny ability to detect things hidden from us — often things of crucial importance such as bombs, drugs, cancer, or lost people — also achieve their amazing feats through the mechanisms of associative learning. Through many months of painstaking and patient practice, a trainer teaches the dog that performing a particular action, often sitting or barking, or both, when the dog notices a critical smell, will be rewarded with the opportunity to chase a ball, or pull on a tug toy, or possibly receive a small food treat.

Whether we are looking at a Westie that seems to know what its owner is going to do next, or at Chaser, who can on command collect any one of many hundreds of objects, or at the thousands of detection dogs, whose names we do not know, that labor daily to keep us safe, there is no shortage of examples of dogs able to do the most amazing things. Yet I don't believe this is actually evidence of anything exceptional about dogs' intelligence. Chaser is remarkable for her work ethic and her powerful relationship with John Pilley. I don't doubt that the gentleman whose Westie appeared to read his mind was also living in the midst of a powerful emotional connection with his dog. Mostly what makes these feats of canine intelligence possible is the relationship between dog and owner and the dog's willingness and enthusiasm to be instructed by this person. This isn't really a unique kind of intelligence. Other animals can be trained to do similar things — and in some cases, even more remarkable ones — if someone has the patience to train them.

Brian Hare was certainly onto something when he said that dogs have a form of genius. A dog that lives as a pet in a warm human household, where it depends at all times on its humans to gain access to everything that it needs — food, water, shelter, the opportunity to enjoy bathroom breaks without risk of chastisement — will become exquisitely and delightfully sensitive to the implications of human actions. This is totally undeniable. Many of us see this in daily life, when your dog seems to read your mind because he knows that you are getting up to make a coffee or take him for a walk. Certainly dogs can do this, and it is a key component of what makes our lives together so successful and satisfying.

The research my students and I have carried out, however, makes it clear that dogs learn the meanings of the things we do because they live with us, not because of any innate, exceptional "genius" for understanding people. The ways we move and act allow dogs to predict what we are going to do next, and so they learn to read meaning into our behavior. They are not born doing this; indeed, dogs living at shelters do not reliably do so, though they can learn quickly. Furthermore, other animals can learn to do this too. The list of animal species that can follow human intentions now includes other domesticated species such as horses and goats, as well as animals that have never been subject to domestication, such as dolphins. I was recently talking to some researchers in Sweden who had hand-reared a number of fallow deer. Knowing my interest in these matters, they were excited to tell me that their fallow deer now follow human pointing gestures.

Given all of this, it is clear that what we see in our own dogs is not great smarts, but rather the result of a phenomenal bond between human and dog. The intensity of that bond makes it possible for dogs and their humans to work together very closely — and, in the case of very patient people and certain highly motivated dogs, to produce some completely amazing performances.

But what is the source of that phenomenal bond between dogs and humans in the first place? Though I was no longer convinced, after our studies at Wolf Park and in a local animal shelter, that dogs possess exceptional intelligence, I couldn't shake the feeling that there was something special about dogs. If it wasn't intelligence, what was it?

My work thus far had convinced me that an answer to this question was vital — for dogs, and for the humans who study and care for them.

Our first forays into the world of animal shelters had not been driven by any particular concern for how homeless dogs are treated in our society. Up to this point, I confess, I had been pretty naive about the lives of dogs that are not someone's pet, and had only been nudged toward the shelter by an intellectual curiosity, a desire to better understand the origins of dogs' ability to follow human intentions. But after our work in the shelter, it was no longer possible to maintain such an impassive approach.

I was taken aback by the impoverished lives of shelter dogs. I hadn't realized that many millions of dogs languish, often for months, in facilities that were designed only for brief stays. They live out their days on concrete floors, with only minimal human interaction each day, and precious few opportunities to simply chase a ball or play in some other way; some dogs are quite literally deafened by the incessant barking of their neighbors and suffer a chronic lack of sleep because of the uncomfortable conditions. They suffer in other ways as well. The two US states I know best, Florida and Arizona, both have deeply uncomfortable summers: subtropical sultriness in Florida and ovenlike desert conditions in Arizona. Yet most shelter dogs in those states are not given respite from summer heat with air conditioning, and the heating they are offered in winter is pretty minimal too.

Our investigation into canine cognition was still in its early stages, but it had already yielded important insights into dogs' minds—insights that, I was sure, had the potential to improve, and possibly even save, dogs' lives. For instance, we had been able to show that, although dogs in the shelter did not spontaneously respond to human gestures, they could very quickly be taught to do so. If (as I heartily recommend) your next dog comes from a shelter, you don't need to worry that she will need lessons in order to understand you. Ordinary daily life, in which humans interact with dogs in a variety of complex ways, provides more than enough experience for a dog to pick up on the implications of human actions, whether gestural or verbal. In normal life a dog probably won't learn quite as quickly as the dogs we explicitly trained at the shelter. Following pointing gestures is something they will learn over the first few weeks in a new home, along with a great many other things, such as whether it's OK to jump on the bed and sofa, and not to chase the cat around the dining table.

Our first forays into the animal shelter had given me a taste of the good that our work could do—but it had also made me aware of the void at the heart of canine cognition research, and the urgent need for better information about dogs and what makes them tick. After our first shelter study, I made it my mission not only to understand what makes dogs unique, but also to determine what that uniqueness means for how humans should be caring for them. I owed it to Benji, to Xephos, and to all the other dogs who had enriched my life to find out what it was that set them apart and to use that information to enrich their lives.

2

WHAT MAKES DOGS SPECIAL?

WHEN XEPHOS CAME into my life, I already had begun to see gaps in the prevailing theory that dogs are special because of their intelligence. Xephos quickly turned those gaps into a gaping hole.

My affection for Xephos kicked in almost as soon as we brought her home — but I could tell that (as I have hinted) this lovable little mutt wasn't very smart. Stairs, for instance, posed quite a challenge. The first house she lived in with us had an upper floor, which apparently was quite a novelty for this little shelter dog. She followed me tentatively up the stairs the first time, but then as I went back down, she just stood at the top and cried. Finally, she summoned the courage to attempt a descent. It didn't go well that first time, and she rolled and tumbled down the last part. No harm done; she gradually figured out this strange human construction.

In 2013, the year after we adopted Xephos, I relocated from Florida to Arizona to start the Canine Science Collaboratory at Arizona State University. This research center is devoted to understanding dogs better, and to improving their lives and the lives of the people they live with, by using the tools of behavioral science. Ros, Sam, and I moved to a house in Tempe that we thought Xephos would like. It has no stairs and even has a little doggy door, so a pup doesn't have to ask permission each time she wants to go outside. But true to form, it took Xephos weeks to figure out how to operate the thing—even with my attempts to explain how it worked by opening it, putting treats on it, and showing her the outside world by lifting the flap. She wasn't quick on the uptake.

Leashes were tricky for her too. I guess her previous family had not taken her out for walks on a leash, because she was forever getting tangled in this strange contraption. In her fascination for everything we came across, she kept walking around me, so that the leash ran all the way around my legs. Or she'd walk on the other side of a lamppost from me, not seeming to grasp why we couldn't both move forward under such conditions. It took a good couple of months before we could have a decent perambulation around our neighborhood.

But while Xephos didn't seem particularly quick-witted, she was (and remains) wonderfully affectionate. Her sweet temperament was already apparent when we picked her out at the shelter, and as soon as we brought her home she displayed a generally warm-hearted demeanor toward pretty much everyone she met (guys with beards being the only exception—she hesitates a little with them). Furthermore, I was astonished at how quickly she set to work to convince us that we were special to her. She seldom lets more than a few feet of ground separate her from one of us. She never misses an opportunity to greet us on our return home, and she loves nothing more than lying at our feet or on the sofa or bed right next to us as we relax. Fortunately, we found that Xephos, unlike many millions of dogs, did not get overtly upset when we had to leave her alone in the house, yet her pleasure at our return clearly knew no bounds. She would make a considerable fuss, even when we had been gone for only a few hours. On the rare occasions when we were forced to be away from her for several weeks, she would cry so badly when we returned that it seemed she was in pain. This sort of anguished relief inevitably made us feel terrible for having gone away for so long.

Even if there was nothing remarkable about dogs' intelligence, I remained convinced—and Xephos worked very hard to make sure

I grasped this — that there is indeed something special about dogs. I could spend all day at the office, reading and writing scientific papers about dogs' behavior, poking holes in the scientific literature about dogs' supposedly unique cognitive abilities. Yet when I came home to Xephos, her wild enthusiasm on seeing me again — so great that it was hard to get in the door as she bounced up to kiss me, once or twice even knocking my glasses off — made it impossible not to recognize that there was something quite extraordinary about these animals, something that set them apart from all other creatures.

The more I thought about it, the more it seemed that this extraordinary thing was not intellectual, but emotional. What set Xephos apart from all the other animals with which I had conducted research or spent time, from pigeons to rats to marsupials to wolves, was her terrific emotional connection to the people around her. The affection and excitement that our presence appeared to evoke in her, and her anguish when we couldn't be with her, were probably the defining characteristics of her behavior toward her human companions.

Although Xephos hadn't been part of our lives for long, she already had led me to question some of my most basic convictions as a behavioral scientist. Much of her behavior appeared to be driven by what I could only characterize as intense emotional attachment to humans. Yet the conventional wisdom and underlying principles of my scientific background and training, behaviorism, strongly suggested that this could not be the case.

Behaviorism is really nothing more than the application to psychology of one of the bedrock principles of science. This touchstone, known variously as the law of parsimony or Occam's razor, dates back to the fourteenth-century scholar William of Occam. I once visited the village of Occam (now spelled Ockham), which lies to the southwest of London, hoping to buy a razor that I could then hold up in my classes, to make an abstract principle more concrete. Unfortunately, the village was so quintessentially parsimonious, there was nowhere to buy even a razor — though there was an excellent pub, where I enjoyed a first-class lunch. In any case, Occam's razor is a principle, not a physical object; it states that the simplest explanation for a phenomenon is always to be preferred over others that allow additional unnecessary explanatory processes to sneak in. This idea is a vital heuristic tool, which has proved immensely valuable over the past six centuries, in disciplines from astronomy to zoology.

As a behaviorist, I was determined to find the simplest, most parsimonious explanation for Xephos's ostensibly affectionate behavior. Not wanting to allow my explanations of animal psychology to include things that we could manage just as well without, up to this point I had tended to shy away from talking about emotions in animals. It was true that, when Xephos bounded up to me as I came in the door from a long day at the university, she certainly looked happy to see me. But the parsimonious scientist in me preferred to see her as acting on previously formed associations between my arrival and the appearance of things she found rewarding, such as walks and dinner. To bring something as messy as emotion to bear on this situation would upset the neat equations of my scientific training and seemed to violate the precept of Occam's razor.

I wasn't alone in my skepticism about considering emotions to understand canine psychology. Many scientists interested in animal behavior do not find emotion a helpful concept. The anthrozoologist John Bradshaw and the dog cognition scientist Alexandra Horowitz, for example, both argue that projecting complex emotions such as guilt onto a dog causes confusion and can even lead us to harm our beloved pooches. To take one example: people often chastise their guilty-looking dogs because they perceive the poor animal's expression as an admission of culpability. In reality, a dog's apparently remorseful expression is nothing more than a manifestation of anxiety in response to an obviously angry human — definitely not an admission of responsibility. The guilty-looking dog does not understand that he or she has done wrong, so punishing the misdeed is misguided, pointless, and cruel.

The neuroscientist and psychologist Lisa Feldman Barrett goes even further, arguing that the very concept of emotion—and the words that we use to categorize different emotions—are human creations rooted in our uniquely human language. They are thus dependent on an appreciation of semantics that dogs cannot possibly possess. Our brains construct our emotions based on our bodies' inner physical states from moment to moment, as well as our lived experience, including the experience of hearing people using specific words to describe their own inner physical states. Barrett concedes that animals may experience broad patterns of positive and negative affective responses, something like the basic "feelings" of anger, fear, happiness, and sadness, but she points out that their inability to comprehend these linguistic categories means that we can't say that they experience these specific emotions, per se.

Regardless of whose theory you subscribe to, the experts' consensus seemed clear: animals' emotions are a scientific black box, a terra incognita that we might never be able to fully explore. But I was developing a sneaking suspicion that nothing about Xephos and her relationship with humans made sense unless one viewed her as an emotional being with the capacity to form strong emotional bonds with our species—a capacity that, I was coming to suspect, was unparalleled in the animal kingdom.

Having been so openly skeptical about other researchers' claims that dogs had special forms of intelligence, I knew full well that by developing my own theory about what makes dogs unique, I was taking on an immense burden of proof. If I were to make the claim that dogs have a special capacity for emotional bonds with humans, I would need evidence that would stand up to possibly quite harsh analysis. Some scientists might (not unreasonably) be as skeptical about my views as I had been about others' conclusions.

So I set out looking for data that might lend support to my hypothesis. And as it turned out, I didn't have to look very far.

Although it is true that modern behaviorists avoid talking about emotions in animals, the famous Russian scientist who in a sense founded behaviorism had no such compunction. He had noticed that dogs appear to form strong emotional relationships with people — and, rather than shy away from it, he placed that observation front and center in his studies.

Ivan Petrovich Pavlov is known to every survivor of an introductory psychology class as the guy who proved that dogs salivate when they are expecting food. (In response, the Irish playwright George Bernard Shaw quipped, "Any policeman could tell you that much about a dog.") This phenomenon — which, students are told, Pavlov demonstrated by ringing a bell just before giving his dogs a morsel of food, until the very sound of the bell was enough to make the dogs salivate — was the result of what is known as "classical" or "Pavlovian" conditioning. Essentially a learned association between a neutral signal and a consequence of importance to the animal, classical conditioning is what John Pilley used to get Chaser to memorize all the different names of his twelve hundred toys. It is a crucially important tool in any dog trainer's toolkit, and a fundamental component of dogs' relationships with humans.

Thanks to the oft-repeated stories about his famous dog-drool experiment, Pavlov's reputation has become fairly one-dimensional — but Pavlov himself was a complicated character. For eighty years after his death we really knew nothing about his personality, but recently an excellent biography by Daniel Todes shone a bright light on the great scientist's life and work. Many of Todes's discoveries blow apart a century of mythmaking about Pavlov. For example, Todes discovered that Pavlov never used a bell in any of his experiments ("bell" being a mistranslation of the Russian word for buzzer). But Todes also explains that Pavlov believed his dogs were individuals with emotions and personalities, and he gave the dogs names that captured their idiosyncrasies.

Pavlov's recognition of his dogs' emotions even shaped his famous experiments. Textbooks make much of the fact that Pavlov had a specially designed laboratory building constructed for his research in St. Petersburg. This impressive edifice, which still stands, is known as the "Tower of Silence" for Pavlov's effort to isolate the dogs inside their testing chambers from any disturbances from the outside world. Photographs show Pavlov's dogs being tested in special soundproof compartments, with the experimenter in an adjoining room, behind a double-glazed window. But what may seem like a cold clinical environment was mitigated by the strong emotional connection between Pavlov and his dogs. Todes informs us that, while it is true that Pavlov expected his students to work at one remove from the dogs in this way, the great man himself would sit inside the chamber with the dog. He knew that these animals needed his company to feel relaxed.

Pavlov needed company too. From 1914 until his death in 1936, Pavlov's most important collaborator was Maria Kapitonovna Petrova. She started out as a student but in time became one of Pavlov's most important collaborators, closely involved in a lot of the research on conditioning that assured Pavlov's fame. Although she may now be largely forgotten, her importance was certainly recognized during her lifetime. From Pavlov's retirement in 1935 until her own retirement at the age of sixty-six, she was the director of the laboratory that Pavlov had founded, and in 1946 she received the Stalin Prize for Science.

In addition to being his most significant scientific follower, Petrova was also Pavlov's lover. The two would sit and whisper quietly to each other inside a dog's compartment about their science or other matters. Sometimes the dog would fall asleep as it waited for an experiment to commence, and woe betide the student who, not realizing that a study was supposed to be in progress, barged in on Pavlov and Petrova as they conversed intimately.

Pavlov, ever the biologist, explained all behavior as a reflex, and so he called the need for companionship that he observed in dogs (and himself) the "social reflex." One of just two Americans who studied with Pavlov, W. Horsley Gantt carried out a study on this phenomenon under Pavlov's direction. He applied a sensor to a dog's chest so that he could measure its heart rate. As a person came into the room, the dog's heart rate sped up in anxious anticipation, but if the person petted the dog, the heart rate would go back down as the dog relaxed.

I came across this forgotten aspect of Pavlov's research shortly after beginning my hunt for evidence for my slowly growing ideas about what makes dogs special. Pavlov's findings about dogs' pronounced physical responses to human presence were ancient scientific history, in a way—but they also offered a nice example of the sort of emotional response that I was interested in, and hoping to study. So my former student Erica Feuerbacher, now a professor at Virginia Tech University, and I set about designing a series of studies picking up Pavlov and Gantt's long-forgotten research on how the presence of people impacts dogs. We wanted to know how important it is to dogs to experience the company of a valued human being. In a sense, we aimed to measure the strength of the emotional response to human presence that Pavlov and Gantt had observed in their studies many decades before.

We decided to take a simpler tack than Pavlov and Gantt had. Rather than measure changes in the creature's heart rate, we assessed the dog's behavior directly. Specifically, we would give the dog a choice between human company and something that we suspected would be equally, if not more, desirable: food. In our earliest studies, we presented dogs with a simple choice: would you rather touch a person's hand with your snout in order to receive a small edible treat, or, as reward for the same minimal effort, have the scruff of your neck gently rubbed and be told that you are a "good boy"? It was as simple as it sounds: when the dog touched Erica's right hand with his snout, she either gave him a small treat from her left hand or stroked his neck with both hands and told him he was a good boy. In some studies, Erica alternated two minutes of food reward with two-minute periods of praise; in other studies she gave the dogs a choice between two people, one who gave treats, the other who gave neck rubs.

We started with dogs living in a shelter, who, we reasoned, don't often get affectionate visitors and therefore might be especially impressed by praise and neck rubs. When that didn't work out the way we expected, we tested pet dogs whose owners acted as the experimenters for us. We thought that if someone who really cared about them spoke sweetly to the dogs, the impact of petting might be greater. But we only got the same result over and over: the dogs seemed to prefer treats to petting and praise. All the dogs we tested, whether they were shelter mutts or pampered pooches at home with their special people, always chose the treats over the human attention.

In retrospect, I'm not sure we got the experiment quite right at first. I think Erica and I both enjoy our dogs' company so much—and were so convinced that they reciprocated that feeling—that we failed to realize that, for a dog who is already in the company of a person, additional neck rubbing is not worth as much as a tasty treat, something that is not always accessible.

With time, however, our research became more sophisticated. We found that if we didn't give out quite so many food treats, so the dogs might have to wait a few seconds before receiving a tasty piece of Natural Balance, whereas they could have their necks rubbed immediately, the dogs' preferences quickly shifted. They started spending more and more time with the person giving out praise and neck rubs, in preference to the person who was slightly slow now in giving out food treats. In this way, they showed us that the praise of a person was actually quite valuable to them. Given a choice between one person who gave out treats once every fifteen seconds, and a second person who gave out neck rubs and soothing words straightaway, the dogs stuck with the neck-rubbing human rather than the one slightly slower with the treats.

Thinking about this some more, we realized that in many ways, the pleasure of the human's company is already there for the dogs in these studies; the humans are present, whether they are giving out neck rubs or not. On the other hand, the treats are being rationed: they are in pouches and the humans give out only one at a time, at specific points in the experiment. For a dog that enjoys human company, just being close to the person may be enough, and the neck rubbing and gentle words may not add much to the situation. A more meaningful experiment would be to take the human away (just as the dogs did not have treats continuously available) and see what they would make of being given access to a human who matters to them after a break both from their human and from food. Erica and I determined to find a way to conduct exactly this study.

Once we figured out the right structure for the experiment, it wasn't very difficult for us to put it together. Erica enrolled some helpers who have dogs but go out to work without their pooches during the day. Such a group of people is not hard to find—more's the pity. There was only one additional criterion: each participant had to have a home constructed with a garage that led directly into the house.

At the end of the workday, after each dog in our study had been alone for many hours, Erica would set up her experiment in the garage of the house that this lonely animal shared with its owner. By the door that led into the house she put two marks on the floor, equidistant from the door and at equal angles from the point of view of someone looking out the doorway from the house into the garage. Then she attached a rope to the door handle and tasked an assistant with opening the door with the rope, so that the assistant could hide out of view from the dog.

Before the assistant opened the door, Erica placed a bowl of tasty dog food on one of the spots on the floor and positioned the dog's owner on the other. The owner had been away at work for a full eight-hour day, and during that time no food had been available in the house. So the dog had been equally deprived of both important things.

Now we had a nice test. When the assistant opened the door and the dog saw the owner and the food bowl—both at equal distances from where the dog was standing and both inaccessible for the past eight hours—which would the dog choose, its special person or the yummy food?

The assistant opened the door.

Invariably, the dog—who had heard the owner drive up—was practically on top of him or her as soon as the assistant opened the door. You could see a momentary look of confusion pass over the dog's face, as it noticed that nobody was immediately on the other side of the door. But within an instant, it spotted its mistress or master and ran over, tail wagging, posture lowered, possibly jumping up to kiss—all in all, very excited to greet this familiar person at the end of a day spent all alone.

Now, at this point in the test, it is quite likely the poor pooch hadn't actually noticed the bowl of food. From a purely technical point of view, the experiment was flawed because the humans in the study were way bigger than the food bowls. But pretty quickly, as the dog circled around the owner, it noticed the other reward. At first the dog just glanced at it—because, compared to greeting the owner, the food was simply of no importance. Then, sooner or later, the dog would actually trot over to the bowl and sniff the contents—but again, would quickly return to the human. Compared to the dog's owner, the food was simply not valuable.

Each time we performed this experiment, we gave the dog two minutes to make the choice between human and food. We never found that, on the first exposure to this test, the dogs had any real interest in the food at all.

With time, of course, as we repeated this experiment every day for a week, the dogs got wise to what we were up to and started eating more of the food. Each day, as the owner came home, Erica and her assistant set up the two spots on the floor, placed a bowl of food on one and the owner on the other (mixing up the left and right positions, so the dog wouldn't develop a preference for going one way or the other), before telling the assistant to open the door and letting the dog make its choice. After a couple of days like this, the dogs began to recognize what was coming. They continued to greet their human first but developed a pattern of running over to the food bowl and stuffing their cheeks with as much dog chow as possible before rushing back to continue greeting their owner.

Notwithstanding the dogs' gradual shift to grabbing some food while greeting their person, these experiments clearly show that, for dogs, the chance to interact with an important human can be as rewarding as food. Indeed, forced to choose, most dogs prefer to be with their person than fed. Over time, of course, they settle down in the company of their human and will eat—why shouldn't they? That doesn't imply that the human isn't important, just that they don't expect the person to suddenly leave.

All in all, the dogs' behavior over the course of this weeklong experiment was a powerful testament to the strength of their bond with their human—and it also made me think about my relationship with my own dog in a new light. No matter how many times Xephos had greeted me effusively after a long day at work, I still had harbored nagging doubts about whether she was truly excited to see me, or simply thrilled at the prospect of receiving her dinner. Erica's garage-door study had neatly posed an answer to that question. Xephos really was excited to see me; she wasn't acting (solely, at least) out of any ulterior motive.

But what was the cause of Xephos's excitement? I knew that Erica's study, while elegant, showed only *that* Xephos cared—not *why,* or rather, *about what.* In order to point to the answer, we would need an entirely different sort of experiment.

Erica's research was motivated by a desire to understand the connections that dogs share with their people, but her next study demonstrated this in a way she did not at first expect. This time, she gave pet dogs a choice between two people: their owner and a total stranger. If your dog had a choice between you and a stranger, whom do you think she would spend more time with? If you said, "Me," you will be surprised at what Erica found. Erica gave dogs exactly this choice, and she found that, in a familiar environment, dogs actually spend more time with a stranger.

This result may seem surprising — surely your dog thinks you are more important than some random person off the street? — but it is actually very similar to what psychologists who study infants call the "secure base effect," a sign of strong attachment to a parent or primary caretaker.

In the 1960s and 70s, a famous pioneer of infant psychology, Mary Ainsworth, developed a natural yet powerfully informative test of the bond between a child (typically younger than two years of age) and a primary caregiver (usually the mother). The aim of Ainsworth's Strange Situation procedure was to probe a young child's relationship with the mother by putting the child into a mildly challenging scenario.

For this experiment, Ainsworth brought mother and child together to an unfamiliar room. At first the child was free to explore while the mother watched, but then the mother abruptly left the child alone in the room with a stranger. Most young kids were pretty upset to be left in a strange place with an unfamiliar person. The mother soon returned, but then left the child again, this time taking the stranger with her; now the child found itself completely alone. The stranger then reentered the room before the mother at last returned, at which point the test ended.

Ainsworth found that infants' reactions to being left alone and then reunited with their mother changed, depending on the strength of the bond between each mother-and-child pair. Children she defined as "securely attached" to their mother tended to explore freely while the mother was present, using her as a secure base from which to investigate the world. These kids were visibly upset when their mother left but happy when she returned and quickly became calmed at this reunion. Children whom Ainsworth termed "insecurely attached," on the other hand, often appeared indifferent to the departure of their mother and showed little emotion when she returned; some of them also showed distress even before their mother departed from the experimental room, and were clingy and hard to comfort when the woman returned.

Ainsworth's Strange Situation procedure provides a structured way of assessing the strength of the connection between a child and its primary caretaker — though people had long recognized that this was important in a child's life, nobody before Ainsworth had found a way to quantify it. The test has now been deployed on many thousands of children and yielded tremendous insights into the subtleties of the relationship between a child and a primary attachment figure.

The basic architecture of Ainsworth's experiment can be easily repurposed to study the relationship between dogs and their primary human figure. In one of the earliest studies of the new wave of research into the relationship between dog and human, József Topál, one of Ádám Miklósi's collaborators at the Family Dog Project at Eötvös Loránd University in Budapest, led a team that investigated how dogs responded to being placed in Ainsworth's Strange Situation. The findings of these Hungarian researchers illuminate the nature of the connection that dogs experience with people. Their findings also help explain why the dogs in Erica's study chose to spend more time with a stranger than with their special human.

Topál and his colleagues tested fifty-one dogs from twenty different breeds, with a fairly even split between males and females. The dogs ranged in age from one to ten years, so they were all adults at the time of the study — but besides the difference in the subjects' species and maturity, the experiment mirrored Ainsworth's original study in almost every respect. Topál's team carried out the Strange Situation test just as it is done with children, allowing two minutes for each phase of the procedure.

Topál found that this test, designed for human children, proved an effective method for assessing the relationship of dogs to their owner. All of the dogs in his study showed evidence of using their owner as a secure base, just as securely attached children did with a parent. Each dog explored and played more when the owner was present than when he or she was gone from the room. When the owner disappeared, the dog was clearly distressed and stood by the door, waiting for the person to return. When the owner came back into the room, the dog was clearly joyous at being reunited, was quick to make physical contact, and initially spent more time with this special human. The researchers concluded that this pattern of behavior, so similar to that of human infants, justified considering the dogs "attached" to their people.

These findings line up very nicely with what Erica had found in Florida, when she discovered that dogs in a familiar environment are more likely to spend time with a stranger than with their owner. And together, these studies suggest that the relationships between dogs and their people are similar to the most stable bonds of attachment between human babies and their parents. Like these securely attached babies, dogs clearly place enormous importance on the presence of their owner. Indeed, when dogs have been deprived of human company for a while, or when they are placed in an unfamiliar location, contact with a familiar person can be an even more important motivator than food.

As I pondered the precise nature of dogs' relationship with humans, I knew this body of research was an important piece of evidence. These studies were revealing a connection between members of two different species that looked very much like attachment. Certainly the patterns of behavior found in these experiments mirrored what psychologists would call attachment when observed between children and parents in our own species.

But what did that attachment imply? My training as a scientific observer of animal behavior had taught me to resist what might seem the natural conclusion here—but my suspicions and theorizing had softened the ground, and I couldn't deny that this early evidence apparently supported my hypothesis: the behavior of the dogs in these experiments implied that they were motivated by an emotional connection with humans.

Excited as I wanted to be by these findings, I resisted the urge to let go and be truly thrilled. There did appear to be evidence for dogs' emotional investment in humans — but what we had so far was only a start. If we wanted to prove anything conclusively — if we wanted to break the law of parsimony — we would need a lot more proof.

Pavlov, Topál, Erica Feuerbacher, and, most especially, Xephos: they all seemed to be trying to tell me that there is an emotional connection between dogs and their people. But I still was reluctant to accept their message at face value. I had been guided to this hypothesis by my instincts as a dog lover, but I was still thinking like a skeptic. I was trying to be careful to rigorously test my theory about the nature of dogs' relationships with humans, even while hoping that it would prove correct.

For one thing, I recognized that living inside a human domicile is actually an option for only a minority of the world's dogs. Perhaps the behavior of these pampered dogs wasn't typical of the species as a whole. Was it somehow caused by living in our homes, almost like children?

Guesses (for that is really all they can be) of the total number of dogs in the world hover a little below one billion. Of that billion dogs, probably something like three hundred million live as pets in people's homes. Many of us live in places like North America, northwestern Europe, and Australasia, where dogs surviving outside human homes are almost (though not quite) nonexistent. But that leaves wide swaths of the globe — including South and Central America, Africa, eastern and southern Europe, and Asia — where many more dogs live outdoors than within four human walls.

If I were to make any claims about dogs as a whole species rather than just certain dogs in particular environments, I needed to investigate the behavior of these un-owned dogs. Yet even that would be tricky. In exploring that question, my colleagues and I would need to find a way of distinguishing dogs that merely *interact* with humans for their own benefit from dogs that genuinely *bond* with people. This subtle but critical distinction had become readily apparent to me during a trip to Russia not long before.

In 2010, as I was passing through Moscow on a research trip, I had a chance to spend a fascinating day with Andrei Poyarkov, a professor at Moscow's A. N. Severtsov Institute of Ecology and Evolution, along with his former student and now collaborator Alexey Vereshchagin, and a few other students. Because he publishes little in English, Poyarkov is much less well known in the West than he certainly deserves to be. I found him to be not just tremendously knowledgeable about dogs, but also a warmhearted man who cares passionately about the canine life in his city. He enthusiastically filled me in on what he had learned over many years of intense study of Moscow's street dogs.

Poyarkov has been studying the stray dogs of Moscow pretty much since they started appearing in significant numbers on the streets of the capital at around the time of the collapse of the Soviet Union, nearly thirty years ago. In the course of a fascinating discussion with him and several students in the research building at the Moscow zoo, I learned a great deal about the plight of these poor animals before and after this watershed moment, and also got a hint of the challenges that humans themselves must have endured in Moscow during the Soviet era. (Me: What happened to stray dogs during Soviet times? Poyarkov: They were rounded up quickly and, if not claimed within forty-eight hours, shot. Cheeky student: Pretty much the same as happened with stray people.)

If you have heard anything at all about street dogs in Moscow, you probably know about the dogs that ride the subway. Certainly that was about the extent of my knowledge of the city's strays before I visited the Russian capital. But though these animals make headlines, they represent only the tiniest sliver of a percentage of Moscow's dogs.

Dogs have very good reasons to be attracted to the metro stations, but not for the trains themselves. At the ground level, these bustling human spaces offer warmth — and the possibility of scavenging tasty leftovers from people who pick up a kebab or a hotdog at a kiosk on their way home from work and then part with a portion of it before descending to catch their train. Also, it's hard for dogs to access the place where trains are boarded (the Moscow subway is exceptionally deep). To go all the way down to the platforms and then board the noisy, careening train cars (the Moscow subway is exceptionally fast) — there's really little benefit to dogs for doing that. Poyarkov estimates there are around thirty-five thousand street dogs in Moscow. He thought just "a handful" rode the trains. Another Russian dog expert, Andrei Neuronov, has counted only twenty that regularly ride the subway. Either way, the numbers make it clear that dogs do not have nearly as much incentive to ride Moscow's subways as they do to roam the streets and stations overhead.

In the evening of the day that I spent at the zoo with Andrei Poyarkov, I went walking around the streets of central Moscow with Poyarkov's collaborator, Alexey Vereshchagin, looking for dogs. Vereshchagin is representative of the new generation of Russian scientists: well-versed in the scientific traditions of his homeland, but also up-to-date on the latest research coming from the West. I was used to the idea of stray dogs outdoors in warm climates, but it was a little strange to see dogs living on streets that, even in September, were getting quite cold. The dogs were bigger than I'd seen elsewhere, and most had thick, shaggy fur that was knotted, matted, and flecked with grime.

At one commuter train station we saw an interesting interaction between three men and a dog. Each guy had a beer bottle in one hand and a hotdog in the other. Judging by how they swayed as they argued, these beers were not their first of the evening. At their feet was a grubby dog of a pretty good size, with long, shaggy hair, mainly white, which may have been naturally interspersed with darker colors or perhaps was flecked with dirt. I didn't want to get close enough to find out. I was keen to hang back and watch how the three men interacted with the dog.

It quickly became clear that these three guys had quite different attitudes toward this animal. One seemed pretty interested in it; he occasionally turned toward the dog and seemed ready to share a piece of his hotdog. But another was dead set against the dog, and whenever it got close enough to him, he made to kick at it. The third guy was completely indifferent; consumed in his own eating and drinking, he didn't appear to even notice the dog.

Observing this scenario, I realized that street dogs probably need to pay even closer attention to people than our house pets do, notwithstanding the fact that pets are very sensitive to us and our actions. Still, in most homes, most of the time, dogs do not have to fear attack, whereas dogs on the street have to be continuously on the lookout for people who may do them harm. It was a poignant reminder of the hardship and uncertainty that roughly 70 percent of the dogs on the planet confront every day.

At another point in central Moscow, Vereshchagin and I came across two dogs lying on the ground near another group of snack kiosks. When we stopped and looked at them for a little while, the dogs started to growl, and when we didn't move off, they got up and left, keeping an eye on us until they were well away. Clearly, for them, people who are not holding anything edible and who get too close are a potential danger worth avoiding.

I could see how these dogs in Moscow were interested in people for food—but did humans mean more to them than a meal ticket? Alas, I was not in Moscow for long enough to investigate that question with Andrei, Alexey, and their team; nor, to my knowledge, have any studies that address this question been carried out in Russia. But fortunately, researchers in other countries have started to fill in this gap.

India is another country with vast numbers of street dogs, and a research group at the Indian Institute of Science Education and Research in Kolkata, led by Anindita Bhadra, has been carrying out some fascinating studies on these homeless animals. Bhadra and her colleagues point out that for many people in India, free-ranging dogs can be a terrible nuisance: they get into garbage and make a real mess of it, and they poop in places where people walk—which, even if the dogs are healthy, creates horrible filth. And most of these street dogs are not healthy. They can be a vector for many serious illnesses, including rabies, which in India still kills around twenty thousand people a year. Most victims catch this truly terrifying disease from dogs. Add to this the dogs' nighttime barking, which disturbs people's rest, and you have an animal that adds up to a substantial irritation for the human population.

Tragically, it is not uncommon for people to kill street dogs in India. Some intentionally poison or even beat them to death, and many dogs are killed unintentionally in road accidents. And yet many people care greatly for these dogs, providing them with food and some form of shelter. From the dogs' perspective, therefore, people must be a very unpredictable commodity. Mother dogs often give birth near or even within human homes, so they must perceive that humans can offer benefits that offset the dangers that our species poses to them. Indian street dogs are also adept at following human actions—successfully passing the pointing test I described in Chapter 1.

Given these facts about the sometimes brutal treatment of Indian street dogs, I would not be surprised if they were at best ambivalent about humans. So I wanted to know—how do they feel toward us, really? Do they fear humans, or are they drawn to us? And if we are important to them, do they have anywhere near the kind of attachment to our species that scientists have detected in dogs that spend their lives being cared for by people?

Testing street dogs is a lot trickier than carrying out experiments on people's pets or shelter dogs, so I was surprised to find a report of a really eye-opening study led by a student in Bhadra's group, Debottam Bhattacharjee, which gets at the question of how street dogs feel about people.

The researchers went out to three locations in and around Kolkata, in West Bengal, and found solitary street dogs. Whereas some free-roaming dogs form groups (people often call them packs, but I tend to avoid that term because these groups are more fluid than the stable packs that wolves form), other street dogs are solitary, and Bhattacharjee had decided to focus on these animals because he wanted to get results from one dog at a time. He and his team gave the dogs a choice between a piece of food on the ground and an identical piece in a person's hand. Not surprisingly, the dogs were wary of the strange people and preferred to eat the food on the ground, but the preference was not very pronounced. Nearly 40 percent of the dogs went up to someone they had never seen before and took the food out of his or her hand.

This result surprised me somewhat, but the next test that Bhattacharjee and his colleagues carried out produced an even more unexpected outcome. In a follow-up study, the team did one of two things to a number of individual street dogs. Either they gave the dog a piece of food, or they patted the dog on its head three times. They did this to each dog a total of six times over the course of a couple of weeks. It really was as simple as it sounds: some dogs were repeatedly fed; other dogs repeatedly petted. Finally, the researchers offered each dog a piece of food and then measured how rapidly the dogs in the two groups approached the person and took the food.

To everyone's surprise, Bhattacharjee and his team found that the dogs who had been petted repeatedly over the preceding two weeks now approached the experimenter more rapidly and were also more willing to take food from that person's hand than the dogs who had been repeatedly fed. In light of these dramatic and unexpected results, the study's authors concluded that "social reward is more effective in building trust between free-ranging dogs and unfamiliar humans than food rewards." Much like the dogs in Erica's study, which were given a choice between greeting their owner and enjoying a bowl of food, the dogs in Bhattacharjee's study appeared to attach greater importance to human interaction.

I did not expect this result, to say the least. To be sure, Xephos had been showing me how important we humans are to her—but from what I had seen of street dogs in Moscow and elsewhere, I had not anticipated that the same would be true of them. I was not expecting social contact with humans to be so rewarding for dogs that live literally as pariahs on the streets of cities where they are often so unwelcome that our species actively seeks them out for destruction. That these Indian street dogs would allow themselves to be petted was in itself somewhat surprising to me, and that this petting would earn their trust better than repeatedly feeding them—this was quite a bombshell. It suggested that positive social contact with our species could be an incredibly powerful thing for dogs, even dogs that did not enjoy a secure attachment with any particular human. Also, it raised the possibility that their "social reflex"—as Pavlov called it—could be a key determinant of their behavior, even more so than their desire for food.

This finding was all the more incredible when you consider that food is a major incentive for this species (as any dog owner will tell you), but an especially big motivator for scrawny mutts living paw-to-mouth on the streets. If you were looking for proof that humans are important to dogs from all walks of life, you couldn't do much better than this.

But while Anindita Bhadra's research group in India had proved that dogs crave "social rewards," they had not speculated as to why this might be the case. Specifically, they had not ventured any suggestions about what it was about socializing with humans that appealed so strongly to dogs. Human contact, clearly, was a form of sustenance for these creatures. But what is it about our presence that they find so nourishing? And was this profound attraction to humans really unique to dogs?

I mentioned before that behaviorists have a reputation, at least partially earned, for ignoring animal emotions. It therefore probably counts as an irony that, as I became more and more interested in proving the precise nature of dogs' unique bonds with people, it was a behaviorist who nudged me further in the right direction.

Mariana Bentosela is a researcher with the National Scientific and Technical Research Council of Argentina, in Buenos Aires. She came and visited with us at the University of Florida for a few weeks, ostensibly to learn some of our research techniques. In point of fact she ended up teaching us much more than we taught her, I think.

Mariana shared our interest in how to characterize the remarkable behavior of dogs: we chatted till late in the night about what makes dogs special and discussed the challenges that each of us was confronting in our research. At the time, I was trying to figure out a quick, super-easy, and reliable way of assessing a dog's level of interest in people. In her studies on how pet and shelter dogs react, given a choice between human company and food, Erica had provided a window onto how dogs feel about us; so had Bhattacharjee, with his study on how Indian street dogs react to people who pet them or offer them food. But these tests were fairly laborious. Was there a simpler way to measure dogs' affinity for people — one that could be employed in the setting where it was needed most?

Aside from our intellectual interest in what makes dogs so remarkable, Mariana and I shared concerns about the welfare of dogs living in shelters — that unpleasant underside to the relationship between our two species. We wondered what might make the difference between dogs that find new homes easily and those that languish in kennels for months or even years, if the shelter does not euthanize them.

Shelter workers and dog-rescue advocates use many kinds of tests to attempt to classify dogs' personalities. In some cases, they want to determine whether certain dogs should be given a chance at adoption; in others, they want to draw more general conclusions about which kinds of dogs are best suited to particular kinds of human homes. But these tests, much like the experiments that Erica and

Bhattacharjee had conducted, are fairly complicated. Mariana had looked into all of them, and wondered if perhaps something simpler might serve to clarify which dogs have the best chance of becoming successful pets.

Mariana and her students in Buenos Aires had developed a wonderfully simple test. Put down a chair in an open empty environment. Mark a one-meter (roughly three-foot) circle around the chair. Have somebody sit in the chair for two minutes, and record what portion of that two-minute interval the dog spends inside the circle. Mariana had already used this test on several dogs back home in Argentina, and she thought it captured well what made the difference between a sociable dog — an easy pet — and a dog that would be more difficult to adopt into a home. As she showed us during several demonstrations of this test in Florida, the sociable dogs spent the vast majority of the time inside the circle with the human, while the unsociable dogs stayed outside of it most of the time.

The author reproducing the dog sociability test with his dog, Xephos

I love simple tests. Simple tests are so much easier to score and so much more difficult to mess up. The role of sitting in the chair was pretty hard to get wrong, and the scoring of how much time the dog spent in the circle was also not exactly rocket science. I could see that Mariana's test held great potential for dogs in shelters. And when I witnessed a demonstration of her simple test on the wolves at Wolf Park in Indiana, I began to appreciate its incredible potential for my own research into dog-human bonds, as well.

Although Mariana had done plenty of research on dogs, before she came to us she had never met a wolf up close. So Monique Udell and I took her along on our next trip to Wolf Park. By the last day of our visit, Monique and I had done all the research we had planned, and so we asked Mariana if there was something she would like to try. Just for the fun of it, she said, "Why not try my simple sociability test?"

Up to this time, I hadn't thought of her easy little test as having any implications for our discussions of what makes dogs special — but as soon as she suggested using it on the wolves, I realized it could provide a very interesting way of evaluating the difference in sociability between these canids and their domesticated cousins. The staff and volunteers at Wolf Park, who are exquisitely well placed to understand the differences between dogs and wolves, had often commented on how — although wolves can be very sweet toward people they know well, even "kissing" people they are keen on — members of this species just don't seem to have open-ended interest in almost everyone, which is typical of dogs. With Mariana's test, we would actually be able to quantify the species' different levels of interest in humans — a truly exciting prospect.

We had a person sit on an upturned bucket in an enclosure with a wolf and, just as Mariana had shown us with dogs, we gave the wolf two minutes to indicate how interested it was in coming within a meter of this person. Just as we had done with dogs, we tried the test with people familiar to the wolves, and also with strangers.

The outcome could hardly have been more dramatic. The wolves at Wolf Park, as I've said, must rank as among the best socialized to human beings you could hope to meet. Many of them can be safely introduced to people they have never met before; these are the ones we tested. These wolves are certainly friendly and also quite noble. In Mariana's test, they didn't try to get away from strange people — nor, happily, did they display any sign of hostility toward the researchers. But these animals displayed very little desire to be close to these unfamiliar humans. The wolves barely ever entered the circle containing a stranger sitting on a bucket.

By contrast, when a familiar person entered the enclosure, the wolves were considerably more interested. They came up to Dana Drenzek, the park director and a young woman they have known all their lives, and spent about a quarter of their time up close with her. The rest of the time, they stayed outside the circle, quietly minding their own business.

Dana Drenzek, the director of Wolf Park, carrying out a sociability test with a wolf

The contrast with what we had found with dogs was striking. The dogs we tested under Mariana's direction spent more time inside the circle with an unfamiliar person than the wolves did with someone they had known all their lives. And when a dog discovered its owner sitting on the chair, it spent every last second close by him or her.

By this point in our studies, Monique and I had made many trips to Wolf Park, and we had always found that other scientists' evidence of differences between dogs and wolves dissolved when we tried to reproduce it. We simply could not replicate their findings. As a result, we had earned ourselves a reputation as the researchers who said there were not really any meaningful differences between dogs and wolves. This wasn't what we believed, of course, but we couldn't deny that, each time we had gone looking for a difference between dogs and wolves that other researchers had discovered, we had been unable to find it.

This time, however, we really had found a difference between dogs and wolves — a massive one. And not a difference in cognition or intelligence, but something much more fundamental: a difference in the animals' interest in getting close to human beings. Something clearly was drawing dogs toward us. The question was, what?

If I have a professional mantra, it is "Proceed with caution." I believe that reliable scientific knowledge can be acquired only by casting a critical eye over even the most plausible-sounding claims. This is especially true when the subject I'm studying is close to my heart — and there aren't many things that are closer to me than dogs, given that I both work with these remarkable creatures and also share my home life with one of them.

When I had been studying rats and pigeons, or even marsupials, fascinating as all those species are and as engaging as they could at times be, there was never any real risk that my personal feelings would overpower my scientific training. But working with dogs, some of whom have made and continue to make powerful claims on my emotions, I was anxious that my feelings might overpower my scientific objectivity.

I had to take a step back and consider how I had got to this point. I thought that dogs' emotional responses to humans might explain their species' powerful and unique connection to our own, and I had a suspicion that it was affection, specifically, that made them behave the way they did. I had uncovered sound scientific evidence that made me think there was something to this theory. But I knew I had only scratched the surface of the revelations that science had to offer, and there was a real risk that, if I dug deeper, I might find that I had been on a fool's errand all along.

On the other hand, I needed to remain open to the possibility that had started me down this road in the first place: that what makes dogs stand out from their wild brethren—perhaps even from all other species on the planet—is their ability to form emotional connections with humans, to feel affection toward us.

I was both distinctly uncomfortable with, yet also intensely curious about, where our research was leading us. I felt myself edging closer and closer to a position that, if not exactly taboo for me, was certainly at odds with my training as a behaviorist. I had been conditioned to seek simple, parsimonious answers to scientific questions. My whole professional life up to this point had been about drawing a bright line between cool, objective scientific descriptions of behavior and warm, fuzzy, but ultimately misleading characterizations of animals as emotional bundles of fur. Yet the mounting evidence for what makes dogs unique among animals—what makes them such unparalleled companions for humans—seemed to be pointing us down a path that came perilously close to what I had been taught to view as soppy mumbo-jumbo.

Emotion seemed to be the crux of the relationship between our species, and dogs' affection for humans seemed to be especially pivotal. This left a behaviorist (and a notorious skeptic) like me in a bit of a bind.

So I did the only thing I knew to do: I kept digging.

3

DOGS CARE

SO MUCH OF DOGS' behavior suggests they are powerfully drawn to humans. I have seen this in parks from Moscow to Tel Aviv, and it was also coming through in the research I had been conducting and reviewing. And this was not just true of pet dogs, pampered daily by their doting masters and mistresses, but was also apparent even in street dogs; they too sought out people, often at the expense of that other cherished reward, food.

But the tests I had considered up to this point only really measured dogs' desire to get *near* humans. The tests hadn't attempted to delve more deeply into what the animals' actions in the presence of their people might tell us about their emotional attachment to human beings. I wanted to answer these questions: How is this emotional attachment to humans manifested in this new layer of behaviors? And does the way that dogs act around people have anything more to teach us about what it is, exactly, that makes them so powerfully attracted to humans?

This was the mystery I turned to next, a puzzle that I sought to solve by looking more closely at dogs' behavior when near people. Happily, once I started to look, I found that other thinkers apparently had been puzzling over this question too, long before I got to it.

One of the earliest scientists to think and write about the relationship between dogs and humanity was Charles Darwin. Like many of us, Darwin loved being near his dogs and was seldom far from one of them. As Emma Townshend recounts in her fascinating book *Darwin's Dogs,* the only period of Darwin's adult life when he did not enjoy the company of at least one loyal canine was the five years he spent on his famous circumnavigation of the globe, on board a boat named, fortuitously, the *Beagle.* (As if its name weren't enough, HMS *Beagle* was a sailing ship classified by the navy as a—wait for it—bark.)

Darwin certainly viewed dogs as emotional beings, inclined to intense feelings toward their human companions. In one of his later works, *The Expression of Emotions in Animals and Man,* Darwin discusses in detail how dogs display that emotion. Early in this book, dismissing those who view emotions as uniquely human, he points out how no living thing can outdo the dog when it comes to indicating emotional connection: "But man himself cannot express love and humility by external signs, so plainly as does a dog, when with drooping ears, hanging lips, flexuous body, and wagging tail, he meets his beloved master."

Darwin goes on to discuss in detail how dogs show affection. He comments on the movement of the tail ("extended and wagged from side to side"), the ears (which "fall down and are drawn somewhat backwards"), and the lowering of the head and whole body. Darwin comments too on dogs' tendency to lick the hands and face of their master. He notes that dogs also lick one another's faces and reports how he has seen dogs licking cats "with whom they were friends." (I think Xephos would rather like to lick the face of our cat, Peppermint, but Peppermint would never tolerate such audacious cross-species fraternization.)

In his descriptions of how dogs show affection, Darwin recognized a profound connection between the behavioral signs of happiness that dogs show in human company and the underlying affection they feel for us. Another of Darwin's important insights was that dogs do not manifest happiness just by wagging the tail; they actually display contentment throughout the body, starting with the face.

Darwin was the first author I know of to consider how dogs' emotions register in their facial expressions – specifically the shape of a happy dog's mouth. What particularly interested Darwin was how happy expressions could be surprisingly similar to angry ones. Thus, he noted that in a happy dog's face, "The upper lip is retracted, as in snarling, so that the canines are exposed, and the ears are drawn backwards." Darwin's theory that the expressions revealing opposite emotions can mirror each other has not stood the test of time as well as his more famous theory of natural selection, but he nonetheless provided a useful impetus to the study of animal emotions.

Happily, though Darwin was the first scientist to study the rich topic of dogs' facial expressions, he was by no means the last. In her fascinating book *For the Love of a Dog*, the renowned dog trainer and behavior expert Patricia McConnell delves deeper into this absorbing phenomenon. She notes that "happy dogs have the same relaxed, open faces as happy people." Surveying photos of people and dogs, she observes, "It's as easy to pick out the happy dogs as the happy people." Her point is a good one; for anyone who has spent any time around dogs, it certainly does feel easy to recognize, by its facial expressions, that a canine is happy.

Whenever I came home and Xephos bounded toward me, it certainly felt as if affection was plastered all over her face. She seemed to be grinning whenever I opened our front door: the corners of her mouth would be raised in what appeared to be a joyful expression, and her lips would be pulled back from her teeth (even if not exactly in a snarling way, with apologies to Darwin).

But how could I be sure that what I was seeing in Xephos's face was indeed a reflection of emotion? Even in the company of excellent guides such as Darwin and McConnell, I remained a tad concerned about reading things into the facial expressions of dogs that are not really there. We recognize that the upturned corners of the mouth of a dolphin, for instance, do not indicate that the dolphin is happily smiling; a dolphin's mouth is just fixed that way; we can see this because its mouth does not change shape, as the human mouth does, in response to the events of daily life. There is nothing to suggest that a dolphin's face offers a window onto its emotions, as our faces do. The expressions on a dog's face, by contrast, certainly do seem to change as life unfolds. Yet how can we know for certain that the upturned corners of a dog's mouth really express happiness, and are not forced upon it by its facial anatomy, as in dolphins, or by other aspects of a dog's characteristic biology?

When I first started thinking about this, I couldn't see how a scientific study into the meaning of dogs' facial expressions could be done. Studies that investigate how people express and perceive emotions involve actors who depict certain emotions; then other people are brought in to rate the actors' expressions. Obviously, actors are trained to express emotions they are not really experiencing, but I couldn't imagine how we could train dogs to do this.

To my surprise, however, one scientific study has found a way around this problem. Tina Bloom and Harris Friedman, from the Pennsylvania Department of Corrections and Walden University, respectively, carried out an experiment to investigate just how well people can identify diverse expressions of emotion on a dog's face. They achieved this by hiring a professional photographer to take photographs of Bloom's police dog, Mal, a five-year-old Belgian Malinois, while he obediently held positions under conditions that most dogs (and many people!) would find very provoking. For example, to elicit a facial expression of disgust, they told Mal to sit and stay, a command that would usually be followed by a food reward; instead, he was offered a distasteful medication. To get a sad photo, they told Mal, "Phooey," a word used during training to let Mal know that he had done something wrong. To elicit a fearful expression, they showed Mal his toenail clippers. For the happy image, Mal's handler told him to sit and stay, and then she said, "Good boy. We are going to play soon." Mal had heard these words many thousands of times as a precursor to an opportunity to play with a ball; thus Bloom and Friedman assumed that hearing them again would put Mal into a happy frame of mind. As soon as the picture was taken, Mal was released from sitting, and the ball was thrown for him. In this way, Bloom and Friedman made a collection of three photographs for each of seven facial expressions, which, in addition to the ones already mentioned, included surprised, angry, and neutral.

Mal, the Malinois in Bloom and Friedman's study, showing (clockwise, from top left) his happy, sad, fearful, and angry faces

Bloom and Friedman then showed these twenty-one photographs to twenty-five people with considerable experience in training dogs, and another group of twenty-five people who had never owned a dog and had minimal exposure to the canine kind. Each person was asked to rate each photograph for the presence of no particular emotion (neutral) or one of six basic emotions: happiness, sadness, disgust, surprise, fear, and anger.

Overall, the human raters were pretty accurate in identifying Mal's emotions, though some pictures were easier to categorize than others. The most difficult emotion to recognize was disgust: only 13 percent of the responses were correct, and people were more likely to perceive Mal's disgusted face as sad. Surprise was also not often correctly identified: only about one of every five responses picked the emotion matching Mal's surprised face. But for all the other pictures, human respondents most often chose the correct emotion. Nearly four out of ten people rightly identified Mal's sad face; nearly half chose the correct answer for the fearful photograph, and seven out of ten recognized Mal's angry face (likely a good thing, for their safety—Mal is a pretty large and powerful dog).

The most successfully identified emotion? Happiness. An impressive nine out of ten respondents rated Mal's happy photos as happy. Recognition was slightly higher for people with a lot of experience around dogs (more than nine out of every ten responses correct) than for people with minimal exposure to dogs (a little more than eight out of every ten responses). But even the lower proportion of correct recognitions still amounts to more than three-quarters of the people who were asked, so it certainly looks like people are pretty good at spotting happy dogs. And what does this happy face look like? It does indeed show a relaxed, gently open mouth curving mildly up at the back—just as Darwin and McConnell described it, and just as Xephos so often demonstrates to me.

Bloom and Friedman's study picks up Darwin's and McConnell's claims that dogs express emotions with their face, and it provides solid empirical evidence that these close observers of dogs were entirely correct—particularly regarding the happiness shown by a dog's smile. This experiment required no expensive, complex equipment, yet it demonstrates that a dog's face clearly can be an accurate window onto what that animal is experiencing and consequently strengthens the chain of evidence for the belief that when our dogs look at us with happy faces, they are experiencing a powerful emotional connection with us. That's good news for those of us who feel that our dogs are happy to be with us, and it adds weight to the case that dogs are experiencing an emotional connection to their people.

Of course, dogs' facial expressions are not the only way that they display their happiness to see us. The tail is another important tool for communicating this pleasure at our presence. Generally speaking, people recognize a happily wagging tail — like a happily smiling canine facial expression — very well. Indeed, I am often amazed by the fact that people readily recognize a dog's wagging tail as an expression of happiness when we have no tail of our own to wag. But as it turns out, a dog's tail holds a few more secrets than its face does, and it can be more difficult for humans to interpret than we might think.

Recently a group of Italian scientists made a detailed study of the wagging tail of a dog and discovered that it possesses dimensions of expressiveness that nobody had ever guessed at. Giorgio Vallortigara and his colleagues at the University of Trieste in Italy found thirty dogs that would be comfortable standing in a black box not much larger than the dog itself while looking out a window at one end of the box. As each dog gazed out the window, Vallortigara's team showed it four different things, one at a time: its owner, an unfamiliar human, an unfamiliar dog, and a cat. While the dog in this box looked at what was presented, video cameras recorded the movement of its tail.

These scientists found that the dogs showed a striking tendency to wag the tail to the right when they saw something they would like to approach — something that made them happy. This rightward wagging was strongest in response to the owner, but was also seen for the unfamiliar human. I'm fascinated to learn that a dog's tail can send such a specific signal of its affection for people — more precise even than many centuries of observation had hinted at. This shows how a dog's affection for us is programmed throughout its body.

Of course, humans aren't the only things that dogs like to approach. The dogs wagged their tails very little when shown the cat, but interestingly, the wagging that was observed was also toward the right. When left alone with nothing to look at and when shown the other dog, the dogs wagged their tails much more toward the left.

Ever since I read this research, I have tried to watch Xephos's tail movements to see whether the findings from Italy match what goes on in Arizona — and I've invited several of my friends to do the same. Unfortunately, it is really too difficult to judge which way a dog's tail is wagging as real life is unfolding around us. Like most dogs I know, Xephos seldom stands still and wags her tail in isolation; she is usually in continuous motion. So I have not been able to confirm, based on Xeph's tail, the results of the study by Vallortigara and his team.

Nevertheless, these findings from Italy add objectivity to the observation that surely many millions of people have made: when your dog sees you, he is happy, and he communicates this by wagging his tail. But Vallortigara's group has also uncovered that there is more to dogs' tail communication than we spontaneously understand. That's the power of the scientific method. If all scientists did was affirm (and occasionally contradict) what laypeople believe about their dogs — that would be a function with some usefulness to it. But to uncover what was previously hidden from human view — in this case, that dogs' tails communicate different things when they wag left or right — that is the true excitement in science.

Knowing how dogs look when they are happy certainly adds to the body of evidence that they experience a connection to their people, because we often see them with happy faces and happy tails when they are with us. But if I was going to come out and claim that dogs are special because of an ability to form emotional bonds with people, I wanted even stronger proof than this.

To be sure, I had by now assembled a considerable body of research. I had found studies reaching back to Pavlov and Gantt in St. Petersburg in the early twentieth century and coming up to the present day in Kolkata, Moscow, Budapest, and North Central Florida; all indicate that dogs have some kind of essential connection to people. And I also had our studies from Wolf Park, which showed that dogs gravitate toward people much more than their closest canine cousins do.

All of these lines of evidence pointed toward a conclusion, but they could also be subject to other interpretations. Our dogs, after all, depend on us for everything that they need — for food, shelter, warmth, and even their toileting needs — so their interest in us could arise just because of the crucial role we play in their lives.

I knew that I needed to do more than simply show that dogs are enthused about people. I needed to show that we *matter* to them. I needed evidence that dogs will actually *do* something to help their humans when we are in distress. That would show that the emotional connection between people and dogs is reciprocated — much stronger support for the notion that dogs are not just attached to us, but also care about us. Evidence of this kind would open up new insights into these animals' emotional lives and would shine light on the canine side of the dog-human relationship.

When I started thinking about the possibility of evidence that dogs might actually do something for their people, I got a terrible sinking feeling. I could clearly remember one of the most vivid conference presentations I ever attended. It addressed exactly this question, and the results were bitterly disappointing.

This was back in the early days of my scientific interest in dogs' behavior — I think around 2004 or 2005 — and I was at the Comparative Cognition Conference in Melbourne, Florida. I attend quite a few scientific meetings, and I have to confess, I sometimes find it hard to maintain my interest through long sessions in crowded rooms. Just between us, I often catch up on sleep in the after-lunch sessions. In one such gathering, however, I was fortunate enough to be awake, and I couldn't believe what I heard.

On this particular afternoon, the speaker was Bill Roberts, from the University of Western Ontario. Not usually one of the more dramatic speakers at a conference, Bill has a dry, laconic style, which may sometimes undersell his always first-rate science. Just as I was settling in for my post-lunch nap, however, I realized that Bill was presenting something rather different from his usual careful laboratory studies on pigeons. Instead, the topic of his presentation had clear and shocking implications for my research into dogs' uniqueness.

Bill was explaining how he had recently conducted a study in which a series of volunteers who were walking their dogs across a chilly Canadian park in November pretended to have a heart attack. Bill's student at the time, Krista MacPherson, was hiding behind a tree with a video camera, and another helper was sitting on a park bench, pretending to read the newspaper. One by one, Bill showed the videos that Krista had captured. One person after another walked into view, stopped suddenly near the "stranger" on the bench, and with a cry of pain and a clutch at the chest, collapsed on the ground. Each dog carefully sniffed its prostrate owner, then responded in one of two ways. Either it lay down next to its master or mistress, or (in the funniest cases) it took two careful laps around the person and, recognizing that nobody was holding the other end of the leash, ran off into the sunset. Not a single dog approached the person on the bench, who might have been able to get medical help.

I have never been at a scientific meeting where there was so much laughter. The impact of seeing these dogs running off, especially after Bill's dry introduction to the research, was absolutely hilarious.

Later, criticism of the study suggested that perhaps the dogs could tell that the owner was only faking a heart attack and not in any real difficulty. Or perhaps the dogs didn't seek help because they didn't know the person on the bench. In response to these criticisms MacPherson and Roberts redesigned the experiment, this time arranging for a bookcase to fall on the dog's owner. They ensured that before this "accident" occurred, the dog was introduced to the "stranger" who could offer help. Krista and Bill even allowed each owner, while pinned to the ground, to explicitly instruct the dogs to go get help. But even with these improvements to the design, the results were exactly the same. As in the heart attack experiment, not a single dog did anything that could possibly help to extricate its master or mistress from the collapsed bookcase.

Several years later, another study was published that further intensified support for the notion that dogs did not seem very inclined to help people. Juliane Bräuer and her colleagues at the Max Planck Institute for Evolutionary Anthropology, in Leipzig, Germany, constructed a compartment, roughly eight and a half by four and a half feet, entirely from Plexiglas. It also had a Plexiglas door, which opened whenever a button on the floor outside the compartment was pressed. Bräuer's group trained twelve dogs to open the door to the compartment by pressing the floor button with their paw. Once they were all reliably able to open the door, the experimenters placed either a piece of dog food or a large key inside the compartment. Since the compartment was fully transparent, the dogs could readily see what was inside before hitting the button. When there was food in the compartment, the dogs almost always pressed the button to open the door — showing that they understood how the door-opening mechanism functioned. When there was a large key on the floor in the room, only about one dog in three opened the door. It didn't make any difference whether a human looked back and forth at the key and the dog, asked the dog to get the key, tried to get the door open by pulling on it, or even said "Open!" in a stern and commanding tone.

In a follow-up experiment, Bräuer's team was able to get the proportion of dogs that would open the door up to 50 percent by pointing straight at the button. However, I presume (and Bräuer and her colleagues support this idea) that the dogs interpreted the pointing gesture as an instruction to press the button — so it still doesn't support the idea that the dogs were interested in helping people.

Both MacPherson and Roberts's and the Leipzig group's experiments are pretty clear evidence *against* the contention that dogs care enough about their people to help them. It also appears that they were very carefully carried out. If you judged by these experiments alone, you would have to conclude that dogs just don't care that much about people.

Fortunately, other studies suggest dogs do have some concern for what happens to people. Ted Ruffman and Zara Morris-Trainor, who work at two universities in New Zealand, had the brilliant idea of exposing dogs to people (or at least the sounds of people) who were in states of extreme emotional anguish, without asking the dogs to do anything in particular—just to see whether the dogs seemed to have any emotional experience whatsoever in response to humans in extreme emotional states.

Ruffman and Morris-Trainor obtained recordings of humans at that most uninhibited phase of life—infancy. No babies were harmed in the making of this science: the experimenters recorded perfectly spontaneous baby cries and laughter. Ruffman and Morris-Trainor set up a pair of loudspeakers and, alternately from each, played recordings of babies crying or laughing. A dog was positioned at an equal distance from each loudspeaker, and each recording was played for twenty seconds at a time. The researchers then measured the dog's tendency to approach one or the other (or neither) of the loudspeakers. Ruffman and Morris-Trainor found that all the dogs they tested were more likely to approach a speaker when it produced the sound of an infant crying.

This finding is intriguing, but it doesn't really tell us all that much. It is perhaps possible that it implies the dogs are concerned about the babies' distress, but the sound of a baby crying might just be a stranger, stronger, or more intriguing noise than a baby laughing; it might be engaging the dogs' curiosity, not their sympathy and concern. However, Deborah Custance and Jennifer Mayer at Goldsmiths, University of London, thought of a way to expand on Ruffman and Morris-Trainor's experiment to make a more telling demonstration of dogs' concern for people.

In designing their study, Custance and Mayer recognized an interesting distinction between empathy and sympathy. Empathy, they argue, is a sort of infection. Seeing you sad makes me feel sad. If empathy is all that I'm experiencing, my response is to seek relief from my sadness. If I were a small child, I might go looking for my mother. (Since I'm not, I might pour myself a Scotch.) Sympathy, on the other hand, is more complex. If I see you are sad and I feel sympathy for you, I am not necessarily sad myself, but I am motivated to try to comfort your sadness. If I were your parent, I might give you a hug. Since I'm not, I might pour you a whiskey. Although it would certainly be interesting if we found that our dogs showed empathy toward our distress, if dogs truly care about their people, then it is sympathy, not empathy, that we should be looking for.

Following on from Ruffman and Morris-Trainor's study, Custance and Mayer also exposed dogs to people in distress but improved the experimental design in several ways. To maximize the chances of getting at the dogs' normal reaction to people in distress, they tested each dog in its own home, and they included the owner as one of the people who would act upset. For twenty seconds at a time, the owner cried as naturally as she could toward her dog. As a control condition, to ensure that whatever the dog did in response to the crying person was not just a reaction to a strange sound emanating from a human, the owner also hummed at the dog for twenty seconds. Mayer—a total stranger to the dogs—also took a turn performing the exact same behaviors. In between the episodes of crying and humming, the owner and Mayer chatted quietly together for two minutes, to give the dog time to relax from whatever reaction the crying or humming had evoked. Both people were present the whole time: all that varied at each step was whether it was the owner or the stranger who made a noise and whether that noise was crying or humming. The order in which the different people and their different behaviors were presented was also randomized for each dog.

If the dogs were just motivated to approach people who were crying because of curiosity at this relatively seldom-heard sound, then they should approach people humming just as readily, since humming was also a rare sound for a human to make around these dogs.

That is not what Custance and Mayer found. In their experiment, the dogs approached the crying people far more than the humming people.

If the dogs were experiencing empathy — that is to say, if seeing and hearing a sad person made them feel sad themselves — then, just like infants who go to their mother whenever they come across a crying person, the dogs should approach their owner for support whenever they hear a person cry, even if that person is not their owner. This is not what Custance and Mayer found either.

Custance and Mayer reported that the dogs in this study went to their owner when she was crying, but they also approached the stranger when she cried. This is consistent with what would be expected from an ability to engage in sympathy — a concern about the welfare of another being and a desire to offer emotional support to whoever is in distress.

To be clear, the conclusion I draw from this experiment is not how Custance and Mayer interpreted their results. They argued that the most likely explanation for their findings was that the dogs, with their extensive experience of living around humans, had probably been rewarded in the past for approaching people who appeared sad.

As I have explained, I appreciate simple, parsimonious explanations for scientific findings such as these. Even if I wasn't able to actually buy a razor in the village of Ockham, I nonetheless believe that the principle of parsimony — limiting the number of explanatory principles one calls upon — is the essential core of scientific explanation. But in this particular case, I am not convinced that Custance and Mayer's reductionist hypothesis is correct. Is it really true that dogs receive more rewards from sad people than from happy ones? I've never tested this idea, but personally, I am more inclined to give a treat to my own Xephos when I am in a joyful mood than when I am downcast. I don't think I'm particularly exceptional in that, and I don't think it's a stretch to say that happiness inspires generosity more than sadness does.

Further, if it is the expectation of reward that prompts dogs to approach crying people, why did they approach the crying stranger instead of their owner? Remember, both people were present at all stages of the test. I would have thought that, if crying leads to an expectation of food, then that expectation would surely be focused on the owner — who has often given food in the past, rather than the stranger, who has never fed the dog. And yet, when the stranger cried, it was the stranger, not the owner, that the dogs in the experiment approached.

No, surely the best explanation for these fascinating findings is not that dogs expect sad people to give them things, but that they are actually concerned about people who are upset. They approached the crying person — whether it was their owner or a stranger — because they were experiencing empathy or sympathy — they were troubled that the person was in distress. This experiment provides convincing evidence that our dogs care about what happens to us.

This experiment of Custance and Mayer's is just the kind of test I love. Like Mariana Bentosela's dog sociability test, it is so simple to carry out, and yet the outcome is really compelling. It is also straightforward enough that you can try it for yourself if you live with a dog. You need no equipment — just a person who is a stranger to your dog, and yourself. And a sofa or two chairs — though I guess you could do it sitting on the floor if you are lissome enough. You need a place free of any distractions, and then you can just take it in turns, crying or humming for twenty seconds, with two-minute breaks in between. You'll be able to see how your dog's reaction compares to those of Custance and Mayer's dogs. Does your dog care about your distress and about a stranger's distress too? Not all of the dogs Custance and Mayer tested behaved in the same way, so it is perfectly possible that your own dog will show a pattern of reactions different from those I have summarized here. You may learn things about your own pup that may surprise you — hopefully in a positive way.

Many of these studies on how dogs react to humans' distress suggest that they care about us—or at least that people matter enough to dogs that they themselves experience an emotional reaction when we appear to be suffering. But at first glance, experiments showing that dogs do not help their people during an apparent heart attack or entrapment under a bookcase seem to contradict this conclusion. How should we reconcile these apparently opposing findings?

One way to resolve this ostensible contradiction is to look outside the laboratory for examples of dogs helping humans. Of course, some types of dogs routinely assist people: two obvious examples are seeing-eye dogs, which help the blind, and Saint Bernard rescue dogs, which find people buried in alpine snow. But these dogs have been trained to help, so their behavior arguably reflects their trainers' intentions, not their own—and thus their actions can't be used to answer the question of whether dogs as a species are motivated to help people.

And yet there are so very many examples of ordinary dogs performing amazing feats to help people in distress. Now, I certainly believe we must be cautious when interpreting the things people say their dogs do, because our love for them can cloud our judgment and our recollection. At the same time, however, the sheer number of accounts of dogs trying to help when people experience acute, obvious, and unfeigned trauma demands that we take this body of anecdotal evidence seriously.

Some of the clearest examples of dogs helping humans were documented during one of the darkest points in the twentieth century. During World War II, UK newspapers carried several accounts of dogs spontaneously digging their owners out from the rubble of bombed homes. For example, in December 1940, the *Daily Mail* reported, "It was Chum, a 12-year-old Airedale companion, who rescued Marjorie French. Her home was demolished, and she was trapped in the shelter when she saw the paws of the dog digging furiously to release her, dragging her out to safety by her hair."

Here we have a situation in which the owner's distress was authentic, not staged. Her cries of pain were doubtless completely convincing. The actions required of her dog were easy to comprehend, and what he needed to do (dig) was something that comes completely naturally to many of Chum's kind. Surely, under conditions like this, dogs really do come to the aid of their owners. (Chum, by the way, was subsequently awarded the bravery medal of Our Dumb Friends League — a leading British animal welfare organization.)

This is pretty compelling evidence that dogs will help people they care about, and there are lots of other amazing stories like this. But such anecdotes pale in comparison to the findings of a very clever experiment carried out on rats at the University of Chicago.

Full disclosure: I once dated a girl who had a pet rat. The little thing scampered enthusiastically around her apartment, but despite its vivacity I've never thought of rats as particularly social creatures. It turns out that I was wrong; rats actually form strong bonds with one another, and two rats sharing the same cage will become real buddies and allies. Just how powerful their camaraderie can be piqued the interest of researchers — and, in due course, attracted my attention as well.

To measure the strength of fellow feeling between two rats sharing a cage, Peggy Mason and her group first devised a small cylindrical container just large enough for a rat to be squeezed into. It's a fairly aversive experience for a rat to be trapped like this, and the poor little rats cry out with squeals of distress that are too high pitched for people to hear, but are clearly audible to others of their own species. The container is fitted with a door that slides shut in such a way that the rat inside cannot open it for himself, but another rat on the outside can open it for him — if the free rat is inclined to help his trapped compadre. Mason and her colleagues first demonstrated that, if the trapped rat and the rat on the outside are friends from the same home cage, many such rats will open the container to get its companion out. Mason's group has gone on to show that the rats will do this even if there is a container of chocolate in the same arena. The free rat will open both containers and share the chocolate with its buddy.

When I heard about this, I had to believe that, if rats will free another individual that is important to them, dogs surely would too. This could be an ideal test of how much dogs care about their people. I realized that all we would need would be a special human trap, with a latch on the outside that is not difficult for a dog to open, and a person willing to be locked in the trap and able to cry convincingly.

We started out by building what I called a "cardboard coffin" out of grocery boxes held together with gaffer tape. It took three large boxes to construct a casket big enough for someone to crawl into. We left the head end of the box untaped and made a hole in the side of it, big enough for the dog to see what was inside. The dog could push her snout into this hole to slide the box open — if she was so inclined.

Xephos was the first dog to be tested with this high-tech apparatus, and I'm mortified to report that she did not attempt to free me from my tomb as I cried out for help. I'm told (by my wife and by my student Joshua Van Bourg, whose project this was) that she ran around, apparently in great distress, and seemed to be trying to get my wife to help — but open the box she did not. On the other hand, when my wife got in the box and called out for help, Xephos straightaway opened the box and rescued its not-as-helpless-as-she-appeared occupant. Let's just call that particular finding a mixed result.

Since that early experience, Josh has built a far more substantial box and asked many people to crawl inside and call out in distress to their dog. This experiment is still ongoing as of this writing, but he already has clear evidence that a lot of dogs will rescue their owner from a container when the person cries out in distress. Josh also found telling differences between the results of the rat study and his dog study. On the first day the rats were put into the apparatus, Mason and her colleagues found that around 40 percent of them freed their buddy—but it took them on average a whole hour to get around to doing so. Even after a week of daily testing, only about half of the rats were freeing each other, and they still took around twenty minutes to do it. By contrast, with the dogs that Josh is testing, he has found that in a single two-minute test, around one-third of them free their owner.

As far as I know, Josh's study represents the first time a scientist has ever tested whether individuals of one species will help members of another. Not only is this a thrilling frontier of science, but also I think it provides clear evidence that dogs have a strong urge to help their owners. Based on other studies, we know that these dogs feel interest in human companionship—but now we also know that they will go out of their way to help people with whom they share a special bond.

Of course, not all of the dogs in this study—or others like it—acted in a helpful way. But I suspect this was a failure of the experiment, not the dogs themselves. The tests are necessarily short, in the interest of attracting volunteers and keeping their cries for help from sounding overly rehearsed; what's more, not all the owners likely faked their distress in a fully convincing way. These issues surely contributed to the finding that not all the dogs acted helpfully.

I also think that some of the dogs wanted to help but didn't understand what they needed to do. I suspect this may have been what happened in MacPherson and Roberts's experiment in Canada. As amusing as the dogs' behavior may appear, many of those dogs actually might have been distressed when their human appeared to have a heart attack or became pinned under a bookcase—the animals simply did not have a clue about what they were supposed to do in such a situation.

Similarly, some of the dogs in our experiment may not have understood how to open the box to let their person out. These are the inevitable limitations of this sort of behavioral experiment. Although we designed the study to offer the simplest possible way for a dog to show its concern and desire to help, it is clear that it remains an intellectual challenge for many dogs. As you may see, if you try this with your own dog, many dogs are quite puzzled by the situation and unsure of how to proceed.

However, in the video recordings we make of the dogs in this experiment we almost always find that the dog's behavior indicates that it is distressed by the scenario, even if it does not open the container and free its owner. What's more, our experiments show that many dogs do help their humans when they're in distress, if the problem is simple enough for the dog to comprehend and the solution involves a behavior it has readily available. Digging and pulling are things dogs know how to do. If we work within these basic parameters, it seems pretty clear that dogs care about us enough to come to our aid.

Over a century ago, one of the American pioneers of animal psychology, Edward Thorndike (often, like Pavlov, viewed as a founder of behaviorism), made a bitter complaint in one of the first books on animal psychology: "Dogs get lost hundreds of times and no one ever notices it or sends an account of it to a scientific magazine, but let one find his way from Brooklyn to Yonkers and the fact immediately becomes a circulating anecdote."

Thorndike's point is a good one: we are naturally drawn to stories of the exceptional and the wonderful. Sometimes these stories may contain a nugget of truth, but often they are exaggerated and not representative of what animals can do. If our understanding of dogs is to become objective and scientific, conferring great advantages in terms of understanding how to care for them, then we need to develop tests and experiments that can establish, without room for controversy, what dogs can really do.

Consequently, in this chapter, I have avoided fictional or secondhand stories of dogs' affection and concern. In much the same way that tales of Lassie's exploits can carry little scientific weight, it would mean nothing to me that the *Daily Mail* in 1940 reported that a dog had rescued its owner from the rubble of a bombed-out house, were I not able to find a way to experimentally test whether this really is something that dogs will do (without actually bombing anyone's house, obviously). MacPherson and Roberts's demonstration that dogs don't seem likely to attempt to help humans in distress is every bit as important in refining our understanding of our canine companions' relationship with us as are the much more positive findings from my student Josh's study. Without objective evidence, I would even be willing to raise doubts about the meaning of dogs' happy faces and wagging tails (though, I confess, the joyfulness that wagging tails convey would be difficult even for me to question).

I take behavior seriously as an indicator of how animals relate to the world — and there is ample behavioral evidence that dogs really do care about people. Dogs seek us out; they ignore food in order to be with us; they express joy, by means of their tail and face, at our presence; they demonstrate a willingness to help us when we are in need. All of this suggests they have strong emotional connections with people, that we *matter* to dogs on a deeper level than most scientists and experts are comfortable admitting.

But I also recognize that motivations are hard to get at by studying actions alone. How do dogs feel about people? Their behavior may contain clues — but their bodies hold the answer.

4

BODY AND SOUL

XEPHOS SOMETIMES MAKES noises that sound like a cross between a whine and a howl. I jokingly describe them as her attempts at English. Generally, I can understand her well enough despite this language barrier. I know that she loves walks and her human family members, she's ambivalent about our cat, she likes human food better than dog food, and so on. But the fact that she and her brethren cannot tell us directly how they feel effectively places a screen between scientists like me and these furry subjects.

The tools of psychology certainly help us peer through the veil that separates our species. Ingenious experiments enable us to observe the relationships between things that happen in the world (like a special person appearing, or a human gesturing toward an object) and our dogs' behavior (seeking proximity to their person, or following the direction of a pointing gesture toward an object). These and many other tests are certainly informative and have advanced our understanding of dogs a great deal. But it is exceedingly difficult to get at the deep underlying motivations for behavior with these kinds of behavioral studies alone.

Thus, even with a growing body of research at my fingertips—and a growing suspicion that Xephos felt an emotional connection to me—I was approaching the limits of my behaviorist skill set. I was also keenly aware that many other canine psychologists did not share my growing passion for research into animal emotions, and thus were not going to be particularly helpful in pushing the envelope in this regard. Fortunately, however, while psychologists drag their heels, another group of scientists — biologists — are charging full steam ahead.

A number of recent scientific studies are focused on finding the biological underpinnings of dogs' responses to people. These experiments include some of the most intriguing and creative ongoing research projects in canine science at the present time. And this body of research, I knew, might hold exactly the kind of unassailable evidence that I needed to prove dogs are special because they can care about people.

If dogs have an emotional engagement with our species, then we should see evidence of it in their bodies — specifically, in the activation of the biological mechanisms underlying emotion. Scientists today have found a range of neurological, hormonal, cardiac, and other physiological markers that are correlated with specific emotional experiences in humans. The fact that all animals are interrelated through their shared evolutionary history implies that similar activity of these same markers in a nonhuman species may suggest that they are experiencing similar inner states.

If dogs truly care about humans, their affection should be reflected in their bodies. Buried in dogs' biology, the evidence of their uniqueness should be visible once we have the right tools to illuminate it.

When we speak of emotion, we naturally tend to talk of the heart. And for good reason: our emotions quite literally can quicken our pulse. Pavlov and Gantt understood this a full century ago, when they became the first people to stick electrodes onto the chest of a dog and measure how its heartbeat changed as it detected a person entering its chamber. These two scientists were able to deduce that the presence of a familiar person calmed the anxious animal.

This line of research was picked up much more recently by two researchers in Australia: Craig Duncan at the Australian Catholic University and Mia Cobb from Monash University. Together they carried out a beautiful demonstration that captured quite literally how two hearts can beat as one when dogs and their humans are emotionally in sync. The study was performed with the support of a dog food company, and a video is available online (search YouTube for "Pedigree hearts aligned"). I know the scientists involved and have spoken to Mia Cobb about this study. The video may be slick, but the research is real and the results are completely compelling.

Duncan and Cobb wired up three people and their pooches with heart-rate monitors. These devices detect not just how fast someone's heart is beating, but also, when two individuals are being recorded simultaneously, whether their hearts beat in synchrony.

For this study Duncan and Cobb chose three individuals who had particularly strong interdependence with their dogs. Glenn is a builder who had suffered massive injuries when the scaffolding he was working on collapsed at a building site; he said he went down a "dark road" after his accident, and he credited his dog, Lyric, with giving him back the will to live. Alice, for her part, has been deaf since birth; her dog, Juno, acts as her ears, giving her awareness of things going on around her that she wouldn't otherwise be able to perceive. And Sienna is a young woman who was devastated by the death of her dog, Max. She didn't think any dog would ever mean as much to her as Max, but her new dog, Jake, was determined to prove her wrong.

The researchers simply asked each human participant in turn to sit on a couch; they attached a heart-rate monitor to the person's chest. In all three tests, Duncan and Cobb could see, from the traces of heartbeats appearing on their computer screens, that their human subject was mildly stressed by the situation. It's an odd feeling, having a probe strapped onto your chest if you are not used to it, and sitting on a couch with cameras on you, capturing your every twitch, is bound to lead to at least a little anxiety. Once each participant was settled, the person's dog, also wearing a heart-rate monitor, was allowed into the room.

As soon as the dog and its person were together, the human's heart rate started to drop, indicating relaxation, and quickly the patterns of heartbeats of human and dog became synchronized: literally, two hearts beating as one. It's as beautiful a demonstration as you could ask for of the intimacy that can exist between human and dog.

This is one experiment you should probably not try at home. Even if you have your own heart-rate monitor, I do not recommend trying to attach one to a dog without professional assistance. Nonetheless, you can find evidence of this same phenomenon in the abiding sensation of calm and deep relaxation that comes from having your dog sit with you. Every person who has enjoyed a loving relationship with a dog has experienced this serene togetherness.

Heart-rate monitors are not particularly expensive or difficult to obtain, so they offer scientists a relatively accessible way to study what goes on in the body of a dog when in a close relationship with a particular human being. But studies by Gregory Berns, at Emory University in Atlanta, Georgia, use a vastly more expensive apparatus. His analysis of the biological bases for how dogs respond to people goes to the core of the organ that controls all our motivations — the brain.

By 2012, Berns was a well-established professor in the relatively new discipline of neuroeconomics, which uses the tools of neuroscience to understand the ways in which humans make economic decisions. Berns and his colleagues train people to lie very still in MRI (magnetic resonance imaging) scanners. These machines use a powerful magnetic field to generate fantastically detailed pictures of the living brain while it is awake. When images of the brain captured while an individual is engaged in different kinds of mental activity are compared, scientists are able to deduce, using a technique known as functional magnetic resonance imaging (fMRI), which areas of the brain are responsible for different domains of thought.

The pictures of brain activity that Berns's team obtain as their subjects puzzle over different kinds of economic problems are so fine-grained, they can identify which centers in the brain are responsible for different aspects of the processing of economic information.

Since keeping one's head perfectly still is critical to the success of this method, the only individuals whose brains have been scanned in this way have been members of our own species. Berns had always counted dogs as part of his family and was fascinated by the puzzle of what his beloved, affectionate beasts might be thinking as they went about life. But it had never occurred to him that fMRI could be deployed to better understand what goes on in a dog's brain.

Berns had a flash of inspiration, however, when he heard about the raid that killed Osama bin Laden in May 2011. What caught his attention was that the US Navy SEAL team that completed this mission included a dog, specifically a Belgian Malinois. As Berns recounts in his fascinating memoir of how he developed this new line of research, he was struck by a photograph of a military working dog strapped to the chest of a soldier parachuting out of a plane. He was tremendously impressed to see that a dog could be trained to cope with such an extreme environment. The dog (like the soldier) wore an oxygen mask, and the noise from the aircraft had to be extreme — never mind the sensation of falling from such a great height.

The fact that dogs can be trained to perform incredible feats under such extreme conditions fired Berns's imagination, and inspired him to conduct a series of groundbreaking experiments on the canine brain — studies that would end up providing powerful evidence to support the theory that what makes dogs special is their emotional engagement with humans.

Berns had taken his most recent dog, Callie, to puppy-training classes with an expert trainer, Mark Spivak. Now Berns asked Spivak whether he thought a dog could be trained to lie still in a brain scanner.

Berns knew that to obtain fMRI data on dogs' brains, he would need cooperation from the animals. For the MRI machine to produce a detailed image of what is happening inside an individual's head, a patient or research subject has to remain perfectly still in the noisy, constraining scanner. MRI machines tend to induce anxiety in people who have received an explanation of the procedure's harmlessness. How should a dog, which can't be reassured by an explanation, be induced to lie perfectly still in such an uncomfortable setting?

Spivak quickly convinced Berns that this could be done. With modern, humane training methods, he argued, it should be possible to get dogs to lie still enough to be studied in the brain scanner.

Berns and Spivak formed a collaboration. Together they built a simple wooden frame in which the dogs were trained to lie down, sphinxlike, with their paws at either side of their head. Berns got recordings of the noise made by the MRI machine during a brain scan, and Spivak trained the dogs to wear headphones, through which he played these unsettling sounds until the dogs were perfectly comfortable with a scanner's clicking and whirring.

Berns and Spivak then moved on to building a mockup of the MRI scanner, so that the dogs could get used to being in the scanner's narrow, constraining tunnel. They even moved the mockup onto a table and trained the dogs to walk up steps and lie down inside the wooden scanner frame, where they then heard through headphones the tiresome noises they would encounter inside the real MRI machine. Throughout this lengthy process of carefully shaping the dogs' behavior, Spivak and Berns used only positive reinforcement (food treats) and didn't advance to the next training step until they could tell, from the dogs' behavior, that they were truly comfortable and ready to proceed.

Finally, after several months of training, Spivak and Berns felt that two of the dogs were up to the task of climbing inside the real scanner and remaining motionless within it. During training, these dogs had performed well enough to assure the researchers that they could cope with confinement within the strange tunnel, along with the noise and vibrations that would occur during testing.

Once he and Spivak had the dogs comfortably taking position in the machine, Berns began work on several studies. Berns shares my fascination for how dogs feel about their people, and he was determined to see whether he could identify, in their brains, evidence of dogs' emotional connection to humans. But in order to interpret how dogs' brains respond to the company of their people, he first needed to confirm that his technique could identify brain regions implicated in processing simple, uncontroversial rewards, like food. Only once he knew how a dog's brain responds to unambiguous food-treat rewards would he be able to deduce the rewarding nature of human company based on the dogs' brain activity.

To see which parts of a dog's brain became active when it perceived that a reward was imminent, Berns couldn't just show a dog some food treats while it was in the scanner. The dog might start to fidget and slobber when it saw food, and this wriggling would mess with the attempt to get clear images of brain activity. Instead, Berns and Spivak taught the dogs that one hand signal (the left hand held vertically) implied that they could expect food, and a second hand signal (both hands held horizontally, with fingertips just touching) indicated that no food was on its way. Each of the two dogs was able to hold its head perfectly still long enough, at each signal, to enable Berns and his collaborators to get a good read on which parts of the brain became active.

With this initial study, Berns and his coworkers found that, when dogs are expecting something attractive, their brains function just like people's brains do. Neurons fire in a very specific area of the brain, called the ventral striatum. This area, a subregion of the cluster of neurons called the striatum, plays an important role in the brain's reward system, which in turn is associated with all kinds of behavior. So finding that this area was also active in dogs expecting a reward validated Berns and his colleagues' approach.

This initial experiment involved just two dogs. Given the extensive training required, Berns, Spivak, and their coworkers did not want to invest time and effort in training a more substantial number of dogs unless they were convinced the method was likely to bear fruit. But with this initial result in their pocket, the team progressed to train more dogs — they now have over ninety dogs who will lie perfectly still in the MRI scanner.

Having demonstrated that they could visualize dogs' specific patterns of brain activity in response to food rewards, Berns, Spivak, and their team moved on to the essence of what they were after — evidence in brain activity for dogs' affection for people. They tested twelve dogs, letting each one smell a cloth holding its own body smell, a cloth carrying the odor of a familiar person, as well as cloths with the odors of an unfamiliar person, a familiar dog, and an unfamiliar dog. This time the researchers found that the ventral striatum was primarily activated by the smell of the familiar person — the dog's primary caregiver. This activity in the centers of the brain related to food rewards thus affirms that the dog's brain processes the presence of a beloved human as a highly rewarding situation.

Now, a cynic might say that this activation in the ventral striatum in response to the scent of a dog's owner doesn't really show that the dog is rewarded just by this reminder of the human that it lives with. Rather, this person has fed the dog so many times, the smell of this special human reminds the animal of food, and the reward centers of its brain become activated by that association.

I met Gregory Berns at a conference and explained my unease to him. I even suggested that he try his experiment with some people who had strong relationships with their dogs but for some reason never fed them. He, quite correctly, pointed out that it would be rather difficult to find people who lived with their dogs but never fed them, and reassured me that he had an experiment in progress that, he felt sure, would address my concerns and assuage my uncertainty.

Sure enough, Berns and the team at Emory University came up with an experiment much smarter than the one I had suggested. This study had three stages, so it's a little more complex — but its remarkable payoff makes it well worth examining.

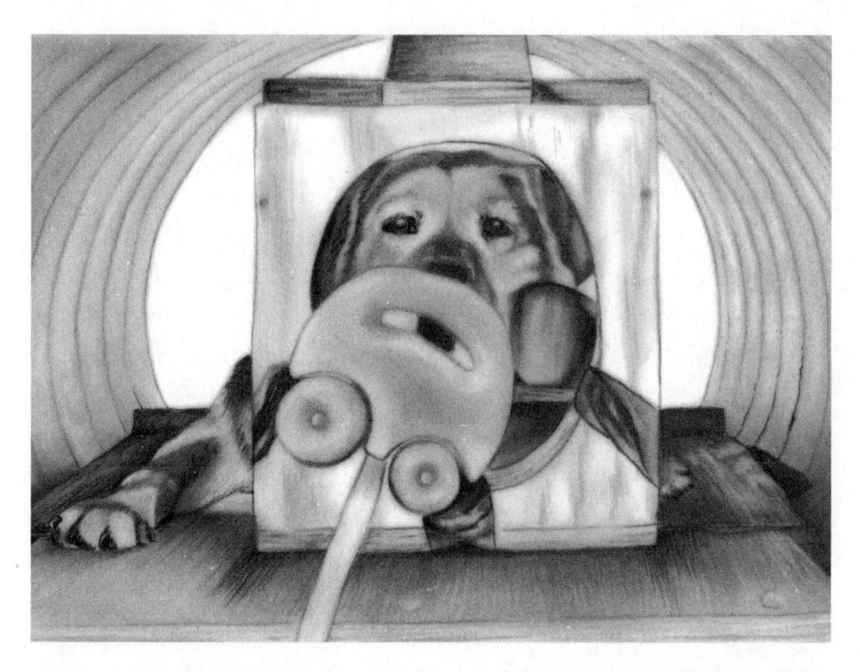

Kady, one of Gregory Berns's subjects, in a mockup of an MRI scanner. The plastic toy car predicts the imminent appearance of her primary caregiver.

First, Berns and his team recruited fifteen dogs and trained them to lie still in the MRI scanner. While each was lying in it, the researchers showed the dog signals indicating that either a food reward or praise from an important human was imminent. If the experimenter showed the dog a plastic toy car on the end of a long stick, that meant three seconds of praise from the owner was on its way. If the dog was shown a plastic toy horse, then a small piece of hotdog would be presented. In this way, for each dog, the scientists obtained a measure of how much the reward centers of the brain were activated by food compared to praise from its special person.

In a second experiment, they repeated this procedure with only the plastic-car praise signal. But this time they occasionally left out the reward (person with praise) that normally appeared after the plastic car was presented. The brain signal for the disappointment when the promised reward fails to materialize is a flipped measure of the brain activity that underlies the happiness induced when the reward does occur.

From these first two experiments, Berns and his team now had two neural signals that indicated how much the dogs valued social rewards from human beings. In the first experiment, when food or social reward was being signaled consistently, they could measure the magnitude of brain activation in the ventral striatum related to human praise and compare it to the level of activation in response to food. The difference between these two signals forms one measure of how much each dog valued human praise compared to food. In the second experiment the contrast between obtaining expected praise compared to being disappointed by its failure to appear is a second measure of the value of human praise.

These two measures turned out to be tightly related. A dog like Pearl, whom Berns described as a "compact and energetic golden retriever," reacted with much stronger brain activation to praise than to food, and also showed the largest difference in brain activity when the signal that indicated forthcoming human praise was not followed by the expected outcome. At the other extreme, a dog like Truffles, whose brain showed little excitement at the delivery of praise (compared to food), also showed little neurological evidence of disappointment when the praise failed to occur. Only two of the fifteen dogs showed more brain activation related to food than to human praise—the brains of the other thirteen were either more activated by praise or showed no difference in activation between these two types of reward.

But the most ingenious part of this study was still to come. In the third experiment the exact same dogs were brought, one by one, into a large room and given a choice between two pathways—one that led to their seated owner, who was ready to offer praise and petting, and another path, which led to a bowl containing dog treats. The seated owner and the bright yellow treat bowl were clearly visible at the point where the dog had to make its choice between the paths.

Each dog was tested twenty times — twenty chances to choose between treats and petting from its owner.

Berns and his colleagues found that most of the dogs showed a preference for praise from the owner rather than the bowl of food, but what they observed went beyond indicating an average preference for petting over treats. Because these same dogs had also been scanned in the MRI machine, Berns's team was able to consider this behavior in light of the dogs' patterns of brain activity. The most exciting finding was that each dog's preference for its owner or the food bowl could be predicted remarkably accurately based on its patterns of brain activation revealed in the scanner. A dog like Pearl, whose brain showed a powerful preference for human social praise over food treats, also, when given a free choice of the two pathways, selected the person giving praise more than twice as often as the food. On the other hand, a dog like Truffles, whose brain was much less responsive to praise (compared to food), also selected food instead of the human, three to one, when given a direct choice between the two. Berns summarized his findings to the *New York Times:* "we conclude that the vast majority of dogs love us at least as much as food." His team's study, however, actually reveals more. It shows not only that many dogs prefer their people to food, but that the area of the dog's brain where the processing of their interest in us takes place is the same region responsible for the analysis of basic rewards like food.

Berns's success in uncovering the mysteries of how a dog's brain processes its preference for people is stunning. Dogs may not speak our language, but through the inventiveness of Berns and his team, their brain can now speak to us directly — and the message is resounding and clear. Dogs' affinity for humans originates from deep within their brain, and their neural activity can even determine how much they care about us. It may not be a stretch to say that dogs are built for affection.

In training dogs to lie still in a brain scanner and observing which parts of their brain become active when they are reminded of their special human, Gregory Berns and his colleagues showed where dogs' emotional investment in humans originates within the overall geography of the brain.

But geography is not the whole story of the brain. Chemistry is another enormously important dimension of cerebral activity. Without chemicals, indeed, we would have no brain function at all. Our nerve cells communicate with one another using dedicated chemicals known as neurotransmitters, and our brain also coordinates the activity of our body through chemical hormones.

The study of these neurochemicals is one of the most exciting frontiers in biology today, helping scientists overcome the language barrier between humans and a range of other species, dogs prime among them. Just as the brain's geography contains powerful clues about how meaningful humans can be for dogs, the chemicals in dogs' brains are also yielding unparalleled insights into the relationship between our two species — and some amazing evidence for how much we really do mean to our dogs.

Recent research shows that a particular hormone plays a starring role in the dog-human relationship. This is oxytocin — a word derived from the Greek, meaning "rapid birth." This substance was first discovered by an Englishman, Sir Henry Hallett Dale, who, in 1909, identified that something in a certain part of the brain causes the uterus to contract. Vincent du Vigneaud (who, despite his rather French name, was an American) won the Nobel Prize in Chemistry in 1955 for identifying that chemical, making it the first peptide (a biological chemical made up of amino acids) to be fully characterized by scientists. Oxytocin is a neuropeptide — meaning that it is a peptide with direct impact on the activity of brain cells.

First implicated in the uniquely female activities of giving birth and producing milk, this crucial peptide is now understood to be present in the bodies of both male and female mammals, and it has a broad role in intimate relationships of all kinds. For instance, when a female rat becomes pregnant, her levels of oxytocin increase, and that change in neuropeptide levels prompts her to take an increased interest in rat pups. By giving virgin rats injections of this mothering peptide, scientists have caused them to become more interested in pups. In sheep too, the release of oxytocin during birth causes the mother ewe to remember the smell of her newborn lamb and thus identify and care for her own offspring and not those of other ewes.

Studies of the crucial role of oxytocin in cementing emotional bonds between individuals are providing the basis for a new and more robust picture of dogs' relationships with their people. They suggest that dogs' affection for humans goes beyond behavior, beyond even brain scans, all the way down to the level of dogs' neurochemistry. The chemicals in dogs' brains, we are now finding, work in concert with their neural geography, orchestrating emotional responses to external stimuli that hold the key to understanding exactly how dogs feel about humans — and where, and how, their affection for us originates in the brain.

A disproportionate part of our understanding of how oxytocin can motivate affectionate behavior comes from the study of a small rodent that lives in the flat land of the midwestern United States and up into central Canada. Prairie voles, unlike other closely related vole species, are generally monogamous, and both parents care for the young. Researchers have found that oxytocin regulates how the prairie vole reacts to the presence and absence of its life partner. For instance, a female prairie vole usually shows a preference for her mate, but she can be nudged to show an interest in an unfamiliar male vole if she is injected with oxytocin when in his company. (A similar effect appears to occur in male prairie voles, although the results are less clear-cut.)

Oxytocin plays a significant role in the way that prairie voles react to their offspring as well as their mates — but so too does the part of the brain that is most responsive to this neurochemical. Researchers first realized this when they noticed that individual female prairie voles differ in how much interest they show in the young of their species, and that this variability is closely related to how many receptors for oxytocin they have in a particular part of their brain, called — you guessed it — the ventral striatum.

The ventral striatum is, of course, the location in the brain that Gregory Berns and colleagues found was energized in dogs by the presence of an important person. This area is a subregion of the cluster of neurons called the striatum, which plays an important role in the brain's reward system, which in turn is associated with many kinds of behavior. It is also, significantly, a region of the brain with a very high density of nerve cells responsive to stimulation by oxytocin.

Taken together, these findings from prairie voles (as well as studies on rats, sheep, and other animals) have given us a clearer understanding than ever before of how specific brain regions and specific brain chemicals work together to cement the social bonds between emotionally connected individuals of the same species. In recent years, researchers have been extending this research to consider the relationship between people and our canine friends.

There is growing evidence that the same neurons and neurochemicals that regulate intraspecies bonds in other animals are responsible for enabling interspecies relationships between dogs and humans. Oxytocin in the ventral striatum seems to play a crucial role in dogs' interest in us — much as it does between a female prairie vole and her mate or offspring. Perhaps the most striking study of this phenomenon comes out of Japan; it focuses on the role of oxytocin in the development and maintenance of emotional bonds.

Takefumi Kikusui and his colleagues at Azabu University in the suburbs of Tokyo are leading the charge in extending our understanding of how oxytocin mediates dogs' response to people. In June 2011 I had the good fortune to visit their research facilities. I have to confess I was quite jealous of their setup. Kikusui has a special building dedicated to dog research, with facilities not just for behavioral study but also for hormonal analysis. But what really made me envious was that they are allowed to bring their dogs from home into the lab with them. When I visited Kikusui and his colleagues in his office, we shared the room with his three standard poodles.

Besides research on the role of oxytocin in lab animals' behavior toward individuals important to them, it is also possible to study the effects of oxytocin in our own species. Some oxytocin gets into the brain when the peptide is sprayed up the nose of a human volunteer, and several scientists have used this trick to manipulate the amount of oxytocin in human subjects' brains, allowing them to observe the results of changes in the levels of this potent neurochemical. For instance, researchers have demonstrated that people experiencing elevated levels of oxytocin are more trusting of strangers. People whose levels of oxytocin have been artificially boosted also remember faces better and are more successful at identifying the emotional expressions shown in pictures of faces. The reason for this seems to be that heightened oxytocin levels prompt people to gaze more toward another person's eyes.

Kikusui's group has combined several research techniques to achieve some really interesting discoveries. For instance, they have taken a urine sample from a human volunteer and have trained each volunteer's dog to pee on command, so that they can analyze the changing levels of oxytocin in the body of the dog and body of the person as they interact. Along with this way of obtaining information about oxytocin levels, they also utilize, on both sides of the dog-human partnership, the painless technique of spraying oxytocin up an individual's nose to manipulate the levels of this neuropeptide. While the person and her dog are interacting, the research group also has video cameras trained on the pair, enabling assessment of how the dog's and the person's behaviors change in concert with their varying levels of oxytocin.

Using this innovative approach to measuring the neurochemical in human and canine subjects, Kikusui and his collaborators have found something incredible: oxytocin levels in humans and their dogs spike when they look into each other's eyes. The magnitude of this effect depends on the strength of the emotional bond between the dog and its owner; owners with a stronger emotional tie to their dogs were found to have dogs who gazed at them for longer periods. This in turn led these people to experience greater increases in oxytocin levels than did owners with weaker ties to their dogs. Kikusui's group has also found that if they used the oxytocin on dogs, the dogs looked at their owners more. When they then measured the levels of oxytocin in the human's urine, they observed that the person's oxytocin levels surged, even though it was the dogs, not the owners, who had been dosed with oxytocin. The research team at Azabu University also found that after receiving oxytocin, the dogs were more inclined to play with people and with other dogs.

These findings beautifully mirror observations of mothers and infants in our own species: mothers with higher levels of oxytocin gaze at their infant's face longer than mothers with lower levels. Any parent, male or female, can identify with the powerful surge of emotion that mothers are surely experiencing in this moment. Just imagine the deep undercurrents of emotion that must be passing between human and canine in similar scenarios, as the exact same elements of their neural machinery are firing furiously.

These findings are tremendously exciting—and they're just the tip of the iceberg. Cutting-edge research into the role of oxytocin in the powerful bond between human and dog is now taking place around the globe. If I were ranking nations' contributions to human-canine oxytocin research, another top contender for the leadership position, besides Japan, would be Sweden.

On a recent visit to Sweden I badly wanted to meet as many of the Swedish oxytocin researchers as I could, but my schedule was constrained by several factors, and one of the young scientists whose work I admired, Therese Rehn, was somewhat tied down with a new baby. Nonetheless we were able to snatch a couple of hours over coffee and tea at Stockholm's central railway station, and we packed a lot into our time-limited *fika* (Swedish coffee break). Rehn was able to add further stunning details about the role of oxytocin in the dog-human bond.

Together with colleagues at the Swedish University of Agricultural Sciences in Uppsala, Rehn has examined the phenomenon that has always struck me as one of the deepest signifiers of our dogs' affection for us: the way they respond when we return home after an absence. When I think about how dogs I have known have shown their feelings toward me, it is their behavior when we are reunified after a separation that comes across as the strongest expression of their affection. Here in Phoenix, Arizona, people are allowed to bring their dogs to the airport. It is such a thrill to see Xephos's excitement when my wife or son appears around the corner from the security melee. I could almost feel envious of how much better Xephos is than I am at illustrating, in her posture and behavior, how thrilled she is to have my wife or son back. My native British reserve makes it hard for me to make a big fuss over people I love in a public place, but Xephos knows no such inhibitions. She cries, she lowers her body and wags her tail low, before jumping up to try to steal a kiss. Strangers turn and stare, but she knows no self-consciousness.

I love the idea of studying what is happening inside dogs' brains in this familiar but mysterious moment—this apparent explosion of affection for a member of another species. For their experiment, Rehn and her colleagues measured oxytocin levels in twelve beagles before and after each one's significant human left the room for twenty-five minutes. The humans were divided into three groups, each with different instructions on how to behave when reuniting with the dog after this brief separation. One-third of the people were instructed to make friendly verbal and physical contact (such as gentle talk and stroking) with the dog, another third were to simply speak to the dog in a pleasant tone of voice, and the final third were told to do nothing but sit passively and read a book. During each reunion, the human and the dog were observed together for four minutes.

Rehn's team found that even when ignored by their human, the dogs registered increased levels of oxytocin at being reunited. However, the more engaged the person was with her dog, the more sustained was the dog's increase in oxytocin. These results demonstrate how dogs' apparently emotional reaction to the reappearance of their special person is indeed underpinned by brain mechanisms that are known to be tied to the most important emotional bonds between individuals.

The other young Swede I was able to chat with while I was in Sweden, Mia Persson, at Linköping University, is taking the investigation of oxytocin's role in the human-dog relationship to an even deeper, more exciting level – down to the very genes that code for the receptors that enable oxytocin's impacts on the brain. By looking at the relationship between dogs' DNA and their proclivity to be more or less excited by human contact, Persson and her collaborators are truly plumbing the depth of dogs' feelings for humans at the deepest possible level of biological analysis.

Persson and her colleagues brought sixty golden retrievers, one by one, into a testing room with its owner. The researchers then sprayed oxytocin up the nose of a dog while it struggled with an impossible problem. For this task, the retriever was offered some clearly visible treats, but the treats were trapped inside a specially constructed container. Quite quickly, a dog in this kind of situation will look imploringly at a nearby human, to elicit help. (This part of the study you can easily try for yourself. It is almost embarrassing how quickly most dogs give up trying to get to something if there is a human they can turn to for help.) Perhaps surprisingly, given what we have seen in the studies from Japan, these oxytocin-spiked dogs did not – on average – look at their human more than the dogs in the control group, which did not receive a spray of the neurochemical.

But Persson's research has an extra layer to it. Using a cotton swab, she and her team took a DNA sample from the inside of each dog's cheek and used this genetic material to analyze the gene responsible for the receptors in the brain that are stimulated by oxytocin. What they found suggests that not all dogs respond to oxytocin in the same way—which helps explain why the intensity of the emotional response to humans varies from one individual canine to the next.

The researchers in Linköping found that the genes that code for the brain's oxytocin receptors are spelled out using just two of the four letters of DNA: A and G. Since every organism possesses two copies of each gene, any individual dog can have an oxytocin-receptor gene spelled in one of the following ways: AA (two copies of A), GG (two copies of G), or AG (one copy of each type). These tiny differences appear to have a big impact on the way that dogs process oxytocin—and how they relate to humans.

Individuals with the first spelling of the oxytocin-receptor gene exhibited markedly more human-oriented behavior than dogs with the second or third spelling. Dogs with the AA version were faster to seek help from their human than were dogs with either of the other two versions, and when oxytocin was squirted up their nose, dogs with the AA type of the oxytocin-receptor gene were even more likely to seek human help.

This mind-blowing finding connects dogs' affection for us to the most basic building blocks of their (and our) biology—the genetic code. The research that Persson and her team carried out, showing us how a particular form of a gene influences the impact of a neuropeptide on the behavior of dogs toward people, connects up the long and winding through-line from the deepest level of biological identity to the high level of behaviors that express emotional states. This is an astonishing thing to pull off—the first step in an exciting new wave of research into dogs' feelings for people.

Researchers have found other intriguing differences in the relationship between dogs' DNA and their relation to humans. Anna Kis and her colleagues at the pioneering Family Dog Project in Budapest, Hungary, for instance, demonstrated the incredible complexity of canine genetics while adding a further fascinating twist to Mia Persson's study of the genes for oxytocin receptors. Kis and her colleagues investigated two different breeds of dog and found different patterns of results. When they squirted oxytocin up the nose of a German shepherd and then a border collie, the outcome depended not directly on the kinds of oxytocin-receptor gene a dog possessed, but on the combination of the form of the gene and the breed of dog. A German shepherd acted in a friendlier manner when oxytocin was squirted up its nose if it had a particular form of the oxytocin-receptor gene. But a border collie was friendlier if it had the form of the gene that led the German shepherd to be less friendly.

This shows just how complex the relationship can be between genes and behavior. The genomes of German shepherds and border collies must be just different enough to cause these subtle differences in their behavioral response to the neuropeptide oxytocin. Specifically, the variances in the genetic code of these two breeds must influence how their oxytocin-receptor genes interact with the neurochemical itself, to produce the patterns of affectionate behavior that we see in these two types of dogs.

From affectionate behavior, to hormones for affection, to genes that code for receptors in the brain for these hormones, scientists are digging deeper and deeper into the biological essence of the dog, finding more and more evidence that its body is programmed for emotional connections. But this evidence, while compelling, does not prove that dogs are unique in this regard. It doesn't answer the question that set me off on this quest in the first place: what it is that makes dogs so special?

As a behavioral scientist, I'm naturally best informed about research into dog behavior. I cannot possibly deny, however, that the defining differences—behavioral or otherwise—between any two species must boil down to each one's DNA.*

If there is something unique about dogs, therefore, I knew that it

---

* Or subspecies: most zoologists today view the dog as a subspecies of wolf, rather than a distinct species of its own.

must be due to their genes. Any abiding, inalienable difference between the behavior patterns of wolves and dogs must be written in genetic code. It may not be easy to find, of course, but it had to be there somewhere. Mia Persson's and Anna Kis's findings offered a hint of how genes influence the characteristic behaviors of dogs, but there had to be more evidence like that out there.

We now know every letter in the big book that spells out the genetic code of the dog. That's because, in 2004, a boxer named Tasha became only the fourth mammal to have its complete genome sequenced, in a project led by Kerstin Lindblad-Toh of the Broad (they pronounce it "bro-ed") Institute, in Cambridge, Massachusetts. This information is proving tremendously useful in helping us understand genetic disease, such as cancer, in dogs. The revelations stemming from this single breakthrough also hold the key to unlocking the mystery of what makes dogs so special.

Five years after the publication of the first dog genome, a young geneticist from the University of California in Los Angeles, Bridgett vonHoldt, led a team that published a paper whose otherwise dry title promised enticing revelations of "a rich history underlying dog domestication." That was enough to get me hooked — and it's no exaggeration to say that what I read in this academic article changed my understanding of what makes dogs the unique beings that they are.

VonHoldt and her colleagues explained how they had gone all the way along the dog genome (actually, the genomes of 912 dogs), comparing it to the wolf genome (actually, the genomes of 225 wolves). The researchers had looked at one little piece of genetic material after another, checking to see whether it showed signs of recent evolution. When we are talking about dogs, "recent evolution" means the process by which certain wolves became dogs — the process commonly called domestication. So vonHoldt and her colleagues were essentially looking for the genetic changes that had made dogs dogs.

I suspect that everyone finds the language of scientific disciplines, other than the one they have been trained in, pretty tough going, but it really seems to me that geneticists take the biscuit. I and my graduate student at the time, Monique Udell (now a professor at Oregon State University), read and reread this paper by vonHoldt and her collaborators. At first we couldn't detect anything that seemed relevant to the question that concerned us—what, at the psychological level, makes dogs special. There was some stuff about genes for "memory formation and/or behavioral sensitization" and some other intriguing bits and pieces, but nothing that seemed to address the question as to whether dogs stand out for their intellect or for their capacity to form emotional bonds.

Then we came across one piece of genetics-ese that popped out at us: the researchers had observed a mutation close to the "gene responsible for Williams-Beuren syndrome in humans . . . which is characterized by social traits such as exceptional gregariousness."

"Exceptional gregariousness"—wasn't that a perfect summation of the phenomenon we were seeing in our behavioral studies? Wasn't that a technical way of talking about the intensely emotional bond that defines dogs' relationships with people? I immediately ran off and did some research on this Williams-Beuren syndrome and quickly discovered that Williams syndrome (as it's commonly known) has many symptoms, but its standout characteristic is greatly exaggerated sociability.

People with Williams syndrome have no notion of "stranger." To them, everyone is a friend. The standard description of a person with Williams syndrome is "outgoing, highly sociable, extremely friendly, endearing, engaging, showing an extreme interest in other people, and unafraid of strangers."

On ABC News online I was able to find a segment from the TV show *20/20* about a summer camp in upstate New York for children with this syndrome. It was dubbed "Where Everybody Wants to Be Your Friend," and the journalist Chris Cuomo was clearly overwhelmed by the warmth of the welcome he received. Undaunted by TV cameras, the kids peppered Cuomo with questions: Where was he from? What was his favorite color? Did he have children? At one point a little girl, perhaps twelve years old, asked him whether he liked girls, and then covered her face in giggling embarrassment as he answered, "I do like girls."

Watching this video, I was immediately reminded of the many comic segments on YouTube in which people pretend to be dogs. I am especially fond of "Cat-friend vs. Dog-friend," by Jimmy Craig and Justin Parker, which has been viewed over twenty-six million times as of this writing. Justin Parker, as the dog, is everything that it is said Williams syndrome children are: he is extremely friendly, endearing, engaging, . . . all the adjectives that come to mind and so well describe the kids on the segment of *20/20*.

I have to admit that I found watching the children with Williams syndrome rather shocking. It may sound absurd, but it felt as if I were watching a whole camp of kids pretending to be dogs. And as soon as I had that thought, I felt ashamed. No matter how much one loves dogs, nobody (I hope) really wants to think of their child as a dog. My own son was, at the time, about the age of some of the children on the TV segment. I wouldn't want anyone to dehumanize him by likening him to a dog.

Emotionally, I was rather discomfited by what I was seeing—but scientifically, I was tremendously excited. The connection between the behavior of kids with Williams syndrome and how dogs behave intuitively felt very strong. Could this be the missing link? The long-sought clue about what makes dogs the remarkable beings that they are?

The more I thought about the scientific implications of what I was seeing, the more I began to feel a certain amount of scientific whiplash as well. In our research comparing the behaviors of dogs and wolves, Monique and I had often pointed out that how an individual behaves is not simply a direct result of a particular genetic heritage. The influence of genes is heavily modulated by life experiences. Back when we had been getting into disputes with other scientists over whether or not wolves can follow human pointing gestures, we had been at pains to explain that doing something like following the gestures made by a member of a different species is not the kind of behavior that springs into the world fully formed when a pup—or a baby—is born. Even our own children are not born following the pointing gestures made by people around them—only some time after their first birthday do kids reliably follow arm points and other bodily gestures. Monique and I were able to show that certain wolves, particular individuals that had received an upbringing around humans that was quite exceptional for a wolf—albeit pretty common for a dog—were indeed ready and willing to follow human pointing gestures and understand their implications.

Having put a lot of effort into emphasizing the crucial importance of experience over genetic identity alone, Monique and I felt a little strange to be so excited about a finding in genetics. But we had never denied the relevance of genetics to understanding how dogs work. And, most obviously, what distinguishes the subspecies of wolves we call dogs from the other subspecies of wolves that are still recognized as wolves must lie in their genetic codes.

Soon after we shared this exciting moment, Monique moved on to establish her own lab at Oregon State University. She and I stayed in touch, of course, and we often talked about our shared scientific fascinations. I think it was within a year of moving to Oregon that Monique told me that she had bumped into Bridgett vonHoldt at a conference. Bridgett had identified the William syndrome genes as a key genetic change in the evolution of dogs from wolves. Monique and I wanted to find a way of testing whether this minute change in the two canids' genomes was the cause of the essential difference in the behaviors of wolves and dogs. We decided to form a collaboration to attack this exciting problem.

Monique, Bridgett, and I needed to figure out a way to tell whether the genes that Bridgett had identified as changed in the journey from wolf to dog were responsible for dogs' "extreme gregariousness," and not some other symptom of Williams syndrome less relevant to the character of dogs. We had to keep in mind that Williams syndrome involves both a large number of genes (about twenty-seven) and that people with the syndrome show a wide range of impacts besides just the gregariousness that intrigued us. Their facial structure is described as "elfin," they may suffer heart problems, their hearing is supersensitive, and they are typically intellectually constrained, among other issues.

It was in the midst of thinking about all this that I had a chance to visit the Wolf Science Center at the Veterinary University of Vienna. For me, what I saw there made the vast behavioral difference between dogs and wolves really concrete.

Founded by the behavioral biologists Kurt Kotrschal, Friederike Range, and Zsófia Virányi, the Wolf Science Center comes about as close as anyone is ever likely to get to rearing dogs and wolves under identical conditions. Located in a lovely village gathered around a castle in the wine country, about an hour southwest of Vienna, the Wolf Science Center contains a population of wolves—around two dozen—that have been hand-reared, so they readily accept human companionship. By the time I visited the center, I had already spent time with hand-reared wolves at Wolf Park in Indiana on numerous occasions, so, although I am always awed by their nobility, it wasn't especially the wolves that I wanted to see. It was the dogs at the center that really intrigued me.

In order to set up the best possible controlled comparisons of wolf and dog behavior, the Wolf Science Center rears and keeps a few dozen dogs under conditions as similar as possible to those of their wolves. That means that the dog pups are taken away from their mother in the first weeks of life and reared by human caretakers. Then, once they are old enough to be independent, they are placed in a fenced area, where they live primarily among just their own kind. Because humans raised them, they happily accept people as social companions, just as pet dogs do. Both the dogs and wolves see and interact with people every day but still lead their lives primarily among their own species. And because the dogs and the wolves have been reared almost identically, any differences the scientists observe between the two species on the psychological tests they administer are surely due to causes other than upbringing.

On the chilly February day I visited, Zsófia, Friederike, and Kurt greeted me and showed me around. The facilities at the center are quite amazing, with multiple enclosures housing wolves and dogs, and a lovely research building that is almost like a clubhouse in the woods.

On our tour of the grounds, we saw the wolves first. They were resting peacefully in the mild sunshine between clumps of snow that remained from a storm earlier in the week. As the wolves heard us approach, many, but not all of them, got up, stretched, and wandered over to the fence. Those in our party who were familiar to the wolves petted them through the wire. Most of the wolves appeared interested and appreciative of the petting. Their tails wagged gently, and they pushed forward to be petted, but they kept their interest in visitors well in check. A few of the wolves completely ignored us.

Then, we walked farther back into the grounds to where the dogs were living. Even before we arrived at their enclosure, the dogs ran toward us, barking, yapping excitedly, and wagging their tails strenuously. The first dogs that noticed us alerted others farther back, and soon there was a cacophony of crazy, excited dogs charging up and down alongside the fence as we got closer.

In this moment, I had to pause and reflect on how different it was to be around these two closely related subspecies of animals. It's hard not to enter a wolf enclosure without, at the back of your mind, some anxiety about your personal safety, even if you know that the wolves at the Wolf Science Center have never harmed anybody. With the dogs, on the other hand, we had no fear for our safety — I was just worried about getting filthy from the snow and mud, because the dogs had such tremendous enthusiasm for jumping up on people.

The contrast between the wolves' mild interest in people and the dogs' frenetic greeting was such a compelling demonstration of how these canid cousins nonetheless show very different levels of attachment to human beings.

I brought home with me this vivid impression of how different these two subspecies of canid can be. It informed my discussions with Monique and Bridgett about how to test the possibility that the genes for Williams syndrome underpinned the very different behaviors of these animals.

Clearly we needed to go beyond the kind of informal impressions of dogs' and wolves' enthusiasm for people that I had gained at the Wolf Science Center. Ideally, we would utilize a simple and quick test that could quantify dogs' and wolves' capacity for cross-species affection in a way that would enable a scientific comparison. This would allow us to say how much of their capacity for affection toward another species the dogs had acquired for themselves, and how much they had inherited from their ancestors, the wolves.

There were a lot of potential tests to consider, but I realized that we had already carried out one excellent task that captured exactly what we needed. Back in Chapter 2, I recounted how our friend Mariana Bentosela from Buenos Aires had introduced Monique and I to what has become one of my favorite experiments. In an open area, she simply has a person sit on a chair, within a circle measuring one meter (about three feet) around the person. She brings a dog in for two minutes and then measures what portion of that time a dog spends within the circle. At Wolf Park, we were able to repeat the test with wolves.

The wolves Mariana tested at Wolf Park, even though they met unfamiliar people most days, showed little inclination to interact with someone they didn't know and spent only about a quarter of each two-minute test period inside the one-meter circle when a human friend was inside the test enclosure. Dogs, on the other hand, spent more time inside the circle with an unfamiliar person than the wolves did with someone they had known all their lives. And if someone familiar to a dog was sitting on the chair, it spent every last second of each two-minute test period close to its human.

Monique and her students also carried out a second, very simple test on the same wolves at Wolf Park and on dogs in Oregon. She gave them a plain plastic food container with a little piece of hotdog inside. To make the task really easy, she passed a thick piece of rope through the lid, so that any beast that wanted to could readily open it. Wolves usually go right ahead and rip the lid off the container to get to the tasty treat inside. But most dogs, if there is a person nearby, would rather look imploringly to the human for help than go straight at the container. This tendency to look across at the person nearby gave us an additional measure of the animal's interest in social contact, this time in a context where the dog or the wolf has a problem it would like to see solved.

After this, we turned to our new geneticist-collaborator for help. Monique sent Bridgett vonHoldt genetic samples (gathered with a cheek swab from inside the mouth) from the dogs and wolves that had undergone these behavioral tests, so that Bridgett could identify whether the differences between dogs and wolves on these tests could be due to the Williams syndrome genes she had earlier identified as subject to recent evolution in dogs.

Although genetics is complex at the level of procedure, conceptually the question we were asking was a simple one, and profound. The dogs and wolves we studied differed in their behavior; they also differed in their genes. Was there a relationship between the different levels of behavioral engagement in our two simple tests of sociability and the genes of the animals we tested?

Although I had my hopes up, I wasn't at all sure that we would be able to find a direct connection between simple behavioral patterns, such as approaching a seated person and looking to a person for help, and the most basic level of biology — the genetic code. Consequently, when Bridgett emailed us to say that the dogs' exaggerated interest in people was linked to three of the genes involved in human Williams syndrome, I was as excited as I had been when, years earlier, Monique and I had found that wolves will follow human pointing gestures. Here we were, showing what it is that makes dogs stand out in nature — the secret of their success with us.

Bridgett was able to demonstrate that one of the genes (the not exactly poetically named WBSCR17) had been under intense selection during the recent evolution of dogs. In other words, it had become altered during domestication. That analysis revealed that for this gene, and two more, known as GTF2I and GTF2IRD1, differing forms of the gene were responsible for the different levels of sociability found in the dogs and the wolves.

As well as this headlining demonstration of a genetic change in dogs on their journey to becoming a species distinct from wolves, the study revealed two further interesting findings. One is that different breeds of dogs possess different versions of these three genes, and the ways that these fall out are consistent with the typical descriptions of breeds as friendly or aloof. Monique and Bridgett are presently carrying out a study of a larger sample of dogs from many breeds to try to get a more precise picture of how genetic variants lead to diverse patterns of sociability in different dogs. A second striking finding is that previous experiments on mice, in which genes have been experimentally manipulated, have demonstrated directly that the GTF2I and GTF2IRD1 genes are involved in sociability. An additional interesting twist is that a minority of people with Williams syndrome do not manifest the exaggerated sociability that is typically a defining aspect of the syndrome. These people, it has been shown, have normal forms of these two genes.

All of this affirms that there is a kinship between our dogs and people with Williams syndrome. New research by Mia Persson and her team in Linköping, Sweden, suggests that other genes, with the even more poetic names BICF2G630798942 and BICF2S23712114, may also play a role in dogs' interest in people. These genes, in humans, are associated with autism. Autism is a syndrome characterized by a reduced — rather than an exaggerated — interest in social contact, but variants of the gene may have different, even opposing effects in dogs. This adds grist to the mill that is grinding hard to find the connections between dogs' genes and their remarkable patterns of behavior.

For all that I was thrilled to be involved with such an exciting scientific breakthrough, I was anxious that parents of children with Williams syndrome might be offended by our discovery that there are genetic similarities between their offspring and dogs. I needn't have worried: the connection made immediate intuitive sense to them. A journalist who reported on our findings interviewed a board member of the United States Williams Syndrome Association. Commenting on these children, she said, "If they had tails, they would wag them."

In the scientific literature, the typical pattern of behavior of people with Williams syndrome is termed "hypersociability" or "extreme gregariousness." This mirrors the careful descriptive language I use in my own scientific writing, where I often employ terms like "affiliation," "contact seeking," or "sociability" to characterize how dogs react to humans. These words label specific behaviors that can be measured objectively. I can observe how a dog cries when it is left alone without a familiar caregiver. I can see the energy a dog puts into greeting a well-known person: his lowered body posture, how he jumps up to try to lick his human on the corners of the mouth. I can measure how a dog acts to console a person who appears upset.

I value the precision of scientific terminology, but I also believe that there comes a time when just labeling and counting individual behaviors becomes obtuse—it amounts to willfully missing larger patterns. For dogs bonded with people, the behaviors and neural and hormonal patterns of response I just mentioned, alongside many others, add up to a bigger picture, and that picture deserves to be called something more than just "sociability" or "gregariousness."

Dogs are not merely sociable; they display actual, bona fide *af-*

fection—what we humans, if we were characterizing it in members of our own species, would commonly call love. The essential thing about dogs, as for people with Williams syndrome, is a desire to form close connections, to have warm personal relationships—to love and be loved.

After witnessing the incredible behavior of the children with Williams syndrome, and after participating in the revolutionary experiments interrelating the unique genetic changes in that syndrome with the affectionate behaviors of dogs, I no longer needed any convincing. Having considered the range of scientific evidence, and having seen the parallels between dogs and humans who share their distinguishing genetic markers, I was comfortable calling a spade a spade. It was only because of this long scientific journey that I felt capable of proclaiming dogs' love for humans—and of doing so with all of the conviction that I had previously brought to my earlier skepticism. I had been as ruthless as possible about interrogating the possibilities that dogs have exceptional intelligence on the one hand and experience affectionate bonds toward people on the other. The challenge I had mounted to both possibilities struck many dog lovers as at best unnecessary and at worst curmudgeonly and mean-spirited. I know this because many people weren't slow to tell me so, from strangers on long plane journeys to many of my best friends. I had often been told to stop worrying and love my dog, and so far as my daily life with Xephos was concerned, that is exactly what I had done.

But there is a payoff to systematic exploration—to striving, so far as possible, to put aside preconceived ideas and collect evidence in an unbiased manner. That payoff is the immense excitement of reaching a conclusion established on solid ground, which can be built upon. With these results concerning the Williams syndrome genes and Mia Persson's findings about the genes for autism, we are at the most fundamental level of the organization of living things—we are looking at the DNA, the code in which life is written. And we can see in dogs' genetic material unmistakable signs of their preparedness to care about us. We can follow this signal back up, through hormones and brain structures, past hearts that beat together as people and their dogs find one another, noting dogs' happy reactions to being with the people they care about and distress at being separated from them, seeing how getting close to their person can sometimes be as rewarding to dogs as the very food they eat, and how they will try to help their people when they are in distress — if they can just understand what needs to be done. At every level of analysis, in studies from independent research groups spread around the world, we see the same message beaming out:

The essence of dog is love.

Love, in turn, is what makes dogs such exceptional — truly, uniquely well-suited companions for humans. Their capacity for love distinguishes dogs from every other animal on the planet, including their closest canid relative, the wolf. Dogs try very hard to get close to and interact affectionately with familiar people, but they are interested in strangers too. In this regard, they are completely different from their wild relatives. Wolves taken from their mother at the earliest possible age and raised entirely by human beings just don't show this level of emotional engagement, even with their surrogate mothers. Wolves can form friendships with human beings, but these relationships never include the all-encompassing love that dogs develop for people.

Today, having arrived at this hard-won understanding of dogs as loving beings, I often feel I am holding something very special in my hands. I know now what makes dogs stand out in the animal kingdom. I have found my professional — and personal — holy grail.

But this knowledge has only made me hunger for more. Specifically, it leads to several crucial new questions that I will turn to in the remaining chapters of this book:

First, how did dogs get to be this way? We now know that their open capacity for loving affection is not shared by their ancestors, the wolves, and this knowledge opens up another great mystery. When, and by what process, did dogs acquire this power to love?

Second, how does love grow in each individual dog's life? I knew from my observations of feral dogs the world over that not all dogs love humans equally, even if they have the power to do so. How does such love develop — and how can we nurture it?

Finally, and most critically, what does the loving nature of dogs mean for these animals and our lives with them? What does possessing this insight, that the essence of dogs is their capacity for love, suggest about the relationships we share with them? Of all the questions I have asked myself, this may well turn out to be the most consequential, urgent, and profound.

5

ORIGINS

LOVE IS THE BIRTHRIGHT of all dogs. But how did they come to possess it — and when?

Accounts of loving behavior in dogs go back to the beginnings of written language. One that hasn't been matched for its emotional intensity was put down nearly two thousand years ago, in ancient Greece, by a man named Arrian of Nicomedia.

Arrian was a philosopher, historian, and soldier who had earned fame for his chronicles of the exploits of Alexander the Great. As a younger man, Arrian had been close to the Roman emperor Hadrian, who plucked him from the ranks of the Roman army for a seat in the imperial Senate. But as he set down his recollections late in life, Arrian's mind was not on Hadrian, or any other of his human friends. Rather, Arrian was thinking about his dog.

Arrian (who also called himself "Xenophon the Athenian" in honor of an even earlier writer on dogs) was writing a book about how to hunt with hounds. Suddenly, in the middle of a section listing the ideal qualities of a hunting dog, he switches to praising Hormé, the dog who rested at his feet as he wrote. Arrian describes how he "reared a hound with the greyest of grey eyes" who was

> most gentle and most fond of humans, and never previously did any other dog long to be with me . . . as she does . . . she escorts

> me to the gymnasium, and sits by while I am exercising, and goes in front as I return, frequently turning round as if to check that I have not left the road somewhere; when she sees I am there she smiles and goes on again in front . . . If she sees us even after a short period of time, she jumps up in the air gently, as if welcoming him, and she gives a bark with the welcome, showing her affection. . . . And so I think that I should not hesitate to write down the name of this dog, for it to survive her even in the future, viz. that Xenophon the Athenian had a dog called Hormé, very fast and very clever and quite out of this world.

This moving tribute by Arrian to his beloved hound not only captures the profound love that humans can feel for dogs; it also beautifully describes how dogs express affection toward people. And it makes clear that dogs' love for people is not some modern affectation, but rather a constant in our relationship with this amazing species, dating back thousands of years.

The roots of this affectionate relationship stretch even farther back in time than this two-millennia-old example. The most ancient written record that I have been able to find that indicates an emotional bond between human and dog is an ancient Egyptian tomb inscription from over four thousand years ago. At just sixty-eight words, this brief record says nothing about how the dog acted toward people, but the very fact that these words were engraved on stone to survive the intervening millennia provides a glimpse at an ancient, affectionate connection between our two species:

> The dog which was the guard of His Majesty. Abuwtiyuw is his name. His Majesty ordered that he be buried, that he be given a coffin from the royal treasury, fine linen in great quantity, incense. His Majesty gave perfumed ointment and [ordered] that a tomb be built for him by the gang of masons. His Majesty did this for him in order that he might be honored.

Linens, incense, perfume, a treasured coffin, a specially built tomb: if you read this epitaph and wonder if your own burial will be half as grand as this dog's, take heart—you are surely not the only one. Over the millennia, this Egyptian ruler's love for his dog has surely impressed the countless people who came across this tomb inscription. But then again, that was the point.

Ancient literature provides many such fragmentary glimpses at a powerful connection between people and their dogs, but written records can take us back only so far. Writing of a sufficient complexity to express thoughts like those that the unidentified Egyptian ruler felt toward Abuwtiyuw (extra credit to anyone who can pronounce that name!) probably didn't exist for more than a few centuries earlier than the time when those hieroglyphs were engraved in stone.

Luckily, we have a great deal of archeological evidence about dogs that predates these written records. Just how far back in time this body of evidence reaches, however, is a matter of intense controversy among archeologists. That's because this evidence is mostly made up of bones—whose secrets can be terribly difficult to decipher. So difficult, in fact, that there is a raging debate in many scientific circles about which bones come from dogs, and which don't.

It might seem simple to differentiate old dog bones from old wolf bones, but in practice, archeological specimens are actually a lot more difficult to differentiate than you might think. The problem is that ancient dogs and ancient wolves were anatomically very similar. Whereas today we think of wolves as big, fearsome animals and dogs as much smaller, gentler creatures, these differences were nowhere near as stark in the long-ago time when the first dogs appeared on the scene.

Early dogs would have been very similar to wolves. We can say this with certainty because it is vanishingly unlikely that all the genetic changes needed to make dogs dogs would have appeared suddenly in one instant package; rather, it would have taken many generations for the two populations of canines to become fully differentiated. This extensive gray area in the evolutionary record makes it devilishly difficult to distinguish ancient dog bones from ancient wolf bones with anywhere near the level of accuracy we need to reconstruct the earliest portions of the history of the dog.

The oldest canine remains that all interested archeologists are willing to agree are definitely from a dog belong to a seven-month-old puppy dated quite precisely to 14,223 years ago (give or take 58 years). These bones were discovered in a quarry near Bonn, Germany, over a century ago, and long forgotten in a drawer in a museum. Only recently have they been carefully analyzed with the latest techniques, and they are now yielding fascinating clues about whether — and how — such early canines could have had loving relationships with humans.

A recent reanalysis of the Bonn puppy hints that its remains may show signs of human concern for the animal's well-being. A team led by Luc Janssens from the University of Leiden in the Netherlands suggested that the puppy suffered from canine distemper, and in order to survive as long as it did, it must have been nursed by people. This is a somewhat controversial conclusion, relying as it does on the ability to interpret markings on the enamel of teeth that have lain in the ground for over fourteen thousand years. If true, however, it would offer a powerful testament to the bond between this long-dead pup and its human caretakers.

Whatever is the case with these bones from Bonn, there is ample evidence that dogs have been loving people for millennia, regardless of whether people reciprocated that love or not. Indeed, I strongly suspect — based on writing by ancient Greeks, ancient Egyptian tomb inscriptions, and many other sources besides — that even people throughout history who may have preferred to steer clear of dogs would have recognized that these animals felt strongly drawn to them. And of course, the earliest written records contain ample evidence that many people did return dogs' love, in spades.

The history of this long relationship is a fascinating one, and — although many of the details remain obscure—it suggests an incredible tale of cross-species love predating the dawn of recorded history. Once I concluded that dogs have the ability to love us, this backstory was at the forefront of my mind. Exactly where did this ability come from? How did relatively aloof wolves, with their natural inclination toward a few intense relationships, morph into dogs, with their contrastingly open attitude to cross-species affection? Where, and how, did dogs' power of love begin?

The journey from wolf to dog took place while there were people around watching—but these people must have had other things on their mind, for they left no indication of how this process unfolded. What's more, the residue that remains leaves a lot of room for guesswork. It is probably precisely because the remnants of dogs' evolutionary journey are so ambiguous that archeologists and geneticists who are interested in the puzzling origin of dogs are so prone to disagreement about how dogs evolved, and what role humans played in the process.

Luckily, in order to understand how the capacity for love first arose in dogs, we do not need to get hung up on the exact date when dogs came into being. What is crucial is the *process* by which dogs evolved, and the role that a great capacity for affection would have played in this evolutionary history.

One version of dogs' origin story—perhaps the most oft-repeated one today—is that dogs arose when our hunter-gatherer ancestors adopted the friendliest wolf pups to help them in the hunt. The eighteenth-century French naturalist Georges Cuvier may have been the first to suggest this model. Over generations, he theorized, selection of the most amiable pups in a litter to be the parents of the next might have gradually created the animal we know today as the dog. This account gains support from the fact that many hunters today find dogs a useful adjunct to their predatory activities; furthermore, some of the earliest representations of dogs show them filling exactly this role.

Ancient dogs' role as hunting companions for humans probably did play an important part in their evolution. What's more, as I'll explain later in this chapter, I believe that dogs' capacity to love people owes a great deal to our time-tested collaboration as hunters. But an experience I had in Israel has led me to doubt that our hunting ancestors can be credited with creating dogs, per se.

In 2012, the same year that my family adopted Xephos, I made a pilgrimage to Israel. Many people visit the Holy Land to see the birthplace of their religion. I was after something different, but perhaps as primal: the origins of the dog.

I had traveled to Israel to see what I believed, at the time, were the earliest remains of a dog. These were the bones of a puppy buried a little less than twelve thousand years ago, alongside a woman, whose hand rests on the pup's belly. This archeological find had fed a belief that dogs arose in the Middle East, and naturally I wanted to see these bones for myself.

I also was keen to see the subspecies of wolf that makes its home in the Middle East: the Arab wolf. I knew this wolf was considerably smaller than the great North American wolves I was already familiar with; it is more the size of a large Labrador retriever. I was particularly curious to know whether this wolf subspecies was any easier to tame than the large gray wolves I already knew a bit about. If they were, this could add heft to the possibility that dogs had arisen in that part of the world.

It wasn't until the final day of my week in Israel that I was able to get up close to some Arab wolves, thanks to a tip from a museum attendant, who told me I ought to visit Kibbutz Afikim, two miles south of the Sea of Galilee. The experience radically changed my views on how dogs might have come into being.

This kibbutz is home to a pair of documentary filmmakers, Yossi Weissler and Moshe Alpert. The museum attendant had encouraged me to visit with them because he knew that Moshe had hand-reared several Arab wolf pups, toward the goal of having them star in a documentary film he and Yossi were working on. The theme: how hunters, thousands of years ago, had engaged tame wolves to help them hunt and thereby started the process that ultimately created dogs.

Unfortunately, Moshe was extremely busy the day I turned up, and I barely had any chance to talk with him. Yossi, on the other hand, had plenty of time to chat. He kindly showed me a four-minute short that he and Moshe had made as a vehicle for drumming up financial backing for their documentary. It was a fairly simple film, but it astonished me. In the movie, a guy in a loincloth, carrying a bow and arrow, goes out hunting with two young wolves. He spots a deer and shoots an arrow at it. The film cuts to the wolves guarding the fallen deer as the hunter catches up, hoists the kill over his shoulders, and sets off for home — the wolves trotting dutifully alongside.

This sounds like a simple enough sequence, but the footage absolutely amazed me. At first I thought I had misunderstood what Yossi had been telling me. Were these perhaps not wolves at all, but dogs (they looked rather like Czechoslovakian wolfdogs)? No, these were indeed the Arab wolves Moshe had hand-raised. Well, then, how had it been possible for the actor playing the ancient hunter to simply lift up the deer in front of the wolves? The wolves I knew from Wolf Park would never have tolerated this removal of dinner right out from under their snouts.

For a moment, I thought I was learning that Arab wolves were a far more tractable subspecies of wolf than the great gray wolves I was used to. If Arab wolves were really this easy to get along with, it might suggest that affection toward people was already present in the particular subspecies of wolf from which dogs possibly descended.

My head was spinning with the far-ranging implications of this video footage — but for better or worse, my confusion was short-lived. Yossi explained to me that the actual filming of the movie had not gone nearly so smoothly as the final product implied. For one thing, Yossi himself was terrified of the wolves. As director he had remained in his car through the whole ordeal, shouting instructions through a slightly opened window.

I should mention that during Israel's wars in the 1960s, Yossi had been a paratrooper. I have always thought that paratroopers are the bravest of all soldiers. I would find jumping out of a plane in flight itself to be adequately terrifying, without the added excitement of shots being fired by people on the ground as I floated defenselessly down to earth. So Yossi clearly was a brave guy, and his fear of the wolves was unlikely to be completely irrational—as he soon explained to me.

Yossi told me that the wolves had actually attacked the actor quite viciously the first time he had reached for the carcass. Filming had to stop for the actor's wounds to be treated. When they returned to reshoot the scene, Moshe held the wolves back while the actor lifted the dead deer.

This fact—that the wolves had refused to share the spoils of the dramatized hunt—aligned far better with how I would expect wolves to behave. It also meant that a movie short, intended to show how wolves could have helped people in the hunt, was actually a very clear demonstration of how using wolves to help with hunting could occur only in fiction.

There was one remaining thing I wanted to clear up. Yossi said that, although he and the rest of the kibbutz were terrified of the wolves, around Moshe and the little kids in his family, the wolves acted in a manner that threatened no danger. Could it be that these wolves were capable of bonding with some people, even while viciously attacking others?

Moshe was terribly busy editing a film up against a deadline. He didn't want to get involved in a Q&A session with this strange academic who had shown up at his kibbutz, but he was willing to meet me and shake my hand. His reddened eyes made it clear that he had been working through the night. He let me ask just one question: "Is it true that the wolves you raised were completely safe around you and your family?"

Silently, Moshe rolled up his right shirtsleeve. Massive streaks of scar tissue provided mute testimony to the fact that not all of his interactions with the wolves he raised had gone smoothly. He didn't have to say anything to tell me everything I wanted to know. Hunting with hand-reared wolves is completely impractical and dangerous. The evolutionary origins of dogs must lie elsewhere.

I can't say I was terribly surprised to find that wolves didn't make good hunting companions. I had been mentally prepared by the late, great Ray Coppinger, a larger-than-life character in the world of dog science and a man who taught me much of what I know about the origins of dogs.

Ray was the first person to poke holes in the idea that dogs had originated as hunters' helpers. He called this theory, pejoratively, the "Pinocchio hypothesis"—not for the puppet's nose, which got longer when he lied (though I'm sure Ray didn't mind this association), but for the earlier part of the story, when the poor craftsman Geppetto creates a marionette, Pinocchio, to alleviate his loneliness.

Together with his wife, Lorna Coppinger, Ray wrote a powerful book called *Dogs: A New Understanding of Canine Origin, Behavior, and Evolution,* which outlined the reasons why dogs could not have been created by humans selecting the friendliest wolves to help them hunt. In their book, the Coppingers enumerated the reasons why this thesis did not deserve to be taken seriously. Their points are still valuable enough to merit a brief summation here.

First, wolves have no motivation to help people hunt. If you tried to go hunting with your pet wolf, almost as soon as you unleashed it, your lupine companion would be miles away, happily filling its stomach while you wandered around the forest, lost and hungry. Hours later, your contented, satiated wolf might come back to you, but you would be absolutely no better off. The wolf would not have brought you any food, nor led you to any prey.

Second, wolves are too dangerous, especially around children, for our ancestors to have tolerated them any more than they had to. To be sure, I have had many friendly, rewarding interactions with hand-reared wolves and have no scars to show for my experiences. But the wolves I met were raised based on a scientific understanding of the type of upbringing likely to instill gentleness and friendliness in these beasts (a topic I will return to in the next chapter). Even then, not all the animals raised in this way can be introduced to unfamiliar people, and they are kept behind twelve-foot-high fences for good reasons.

Third, in order to select friendly wolves for breeding, early humans would have needed far more foresight — and have known far more about genetics — than we can possibly give them credit for. Fourteen thousand years ago (or more), there were no other domesticated animals. People would have no way of knowing that the big, scary carnivores around them could one day turn into friendly and useful companions if they just tried breeding them selectively for a few centuries.

Ray and Lorna Coppinger argued that the earliest dogs did not occupy a niche as hunting companions for humans. Rather, it is far likelier that dogs evolved to fulfill a much more prosaic, even pitiful, role: that of scavengers around early human settlements. The Coppingers pointed out that, when people became settled, they also started producing mounds of trash. This refuse would have attracted (and — despite our best efforts — continues to attract) many species of animals. Certain wolves, they theorized, were among those dumpster divers.

Dogs likely originated in places where our ancestors found exceptionally rich resources for hunting and gathering, and thus settled down for years, or even generations. Humans settled in one place inevitably produce that unique marker of our species: mounds of garbage. In doing so, they created a new opportunity. Trash is material that we as humans view as worthless, but it can be valuable to other species. As Aristotle said, "Nature abhors a vacuum," and bones that humans have stripped bare of meat still contain nutrients that other species can exploit.

To this day, in many parts of the world, diverse species congregate at trash dumps. In Kolkata, India, cattle roam through the city dump; in Alaska, people have to watch out for polar bears scavenging around their trash heaps. Thousands of years ago, wolves must have adopted the same food-finding tactic, sniffing out edible remnants near our ancestors' camps.

Scavenging is a habit that wolves in some parts of the world still carry on, as I had a chance to see firsthand on that same journey to Israel. Earlier in my trip, I had gone to the Negev desert, in the south of the country, to see Arab wolves in the wild. The National Parks ranger who kindly took me out to search for these animals immediately made a beeline for the town dumps, which are scattered around the desert. As the ranger explained, these dumps are the most likely congregating places for wolves in the Negev because that desert environment offers few conglomerations of edible material, and certainly no others as large as the dumps.

Abundant evidence from around the world testifies to the way that wolves are attracted to human dumps. The same goes for dogs —even more so. This canine scavenging would be a more familiar sight if it weren't for first-world governments' investment in fences and dogcatchers to keep these animals out of city dumps. And you don't have to travel far outside these bubbles of prosperity to find dogs on rubbish heaps today, no matter how developed the country. I've seen them in places as diverse as Sicily, the Bahamas, and Moscow. Though none of them qualify as a third-world location, each nevertheless hosts plenty of dogs eking out a living around any rubbish that isn't fenced and guarded.

For any sort of animal, canid or otherwise, the key to profiting from the rubbish that humans create is tolerating—and in turn, being tolerated by—humans. But wolves and dogs, while similar in so many other ways, are extremely different in this regard. Unfortunately, no study has been carried out on the town dumps I visited in

Israel, nor any other location where both dogs and wolves seek out a living on human garbage side by side. Researchers have, however, looked separately at trash-scavenging wolves in Sweden and dogs in Ethiopia. The wolves in Sweden run off when they detect a human within 650 feet. Dogs in Ethiopia let unfamiliar people get to within about 16 feet before moving aside.

That difference in a single measure—what biologists call "flight distance"—adds up to an enormous difference in how much food the two closely related canids can extract from a human dump site. By being more tolerant of—and tolerated by—humans, dogs are able to get much more from human waste sites than wolves can. This ability to tolerate the presence of people is therefore a major adaptive advantage for dogs—in scavenging situations, at least.

I recognize that the idea of dogs coming into being on trash dumps is much less appealing than the alternative story of hunters picking up wolf pups that will help them pursue prey. The journalist Mark Derr, in his account of the origins of dogs, *How the Dog Became the Dog,* became quite carried away with disgust at the idea that our beloved canine companions could have started out as dumpster divers. He wrote with revulsion of the thought that "the wolf [would have] voluntarily [become] a sniveling midden maven, a foul-tempered, slinking, village offal eater—[a] 'diaper cleaner.'" But the truth is, for all that we love to imagine our ancestors as lords and ladies hunting on horseback, most of us have to face up to the fact that we are derived from a long line of peasants, eking a living from recycled remnants. And what's true for us is most likely true for our canine best friends too.

Much as we might like to, we don't get to pick our past. Both we and our canine friends are scavengers of sorts. Perhaps there is something fitting about that common history—that shared modus operandi. Could it be what explains their affection for us? Or do the origins of dogs' love lie elsewhere?

With the available scientific evidence, we have no way of knowing whether the earliest wolf-dogs loved people the way our dogs love us today. But my best guess is that they did not.

I suspect that at the earliest stage of canine evolution, when dogs were basically still wolves (albeit ones that had largely given up hunting large prey and had developed a greater tolerance for human beings in order to feast on our trash dumps), these animals still had mostly wolfy personalities. They likely were inclined to a small number of strong bonds, almost always with members of their own species; these proto-dogs would not, in other words, have been the promiscuous social creatures that our canine best friends are today.

That is not to say that our ancestors would not have noticed that these animals were different from wolves. These proto-dogs may well have evoked less fear in their human neighbors than "real" wolves did. Since they were no longer as keen on hunting live prey, the first dogs probably were less fierce and formidable. They likely developed smaller and less powerful jaws and teeth, and their behavioral development may have started to slow down, so that as adults they retained juvenile behaviors such as playing and forming friendships. They may have huffed and gruffed—the forerunner sounds of barking, which is rare in wolves—when animals they were afraid of (such as bears and "real" wolves) approached the camp. These warning vocalizations may have made them somewhat useful to their human hosts.

But those distinctions aside, I'm not inclined to see these animals as the full-on love machines whose company we enjoy in our homes today. At least, I will resist that conclusion until science shows me otherwise—which it certainly might, in the not-so-distant future.

In order to figure out when dogs became the hypersocial, intensely loving beings they are today, scientists will need to identify at which point in their evolutionary history dogs' genomes mutated to include the Williams syndrome genes (described in Chapter 4). Right now, my friend and collaborator at Oxford University, the zooarcheologist and geneticist Greger Larson, is searching archeological remains of early dogs for signs of these genes. He may produce an answer for us at any moment. If and when he does, he will shine a light deep into our two species' intertwined histories, spotlighting the precise moment when humans and dogs fell in love — or at least when dogs' love for humans began to invite similar feelings on our part. In the meantime, we must content ourselves with informed but unverifiable speculation.

Personally, I believe that dogs acquired the power to love not during the earliest, scavenging phase of their species' history, but rather at a more recent phase in their evolutionary journey. The critical shift, I think, would have come when their ancestors and ours set off from the settlements where these animals had been scavenging, and embarked together on a hunt.

As I've explained, wolves cannot be viable hunting companions for humans — but these newly human-tolerant canids were not wolves. They would not have had the same aggressive tendencies as wolves, and probably were not as good at hunting independently (which is one of the traits that make wolves such poor hunting partners for people). What's more, they would have evolved to be more tolerant of humans at a crucial moment in our own species' history: a time when we especially needed dogs' help.

Because scientists now know that dogs arose at least fourteen thousand years ago (and some archeologists believe it was even much earlier than that), we also can be quite certain that dogs came into being during the last ice age. After blanketing the planet for many tens of thousands of years, the ice sheets started to disappear about twelve thousand years ago. It is clear that dogs originated somewhere within this icy time frame.

As you can imagine, this millennia-long cold snap put unique pressures on the humans who lived back then. But they were well accustomed to this climate by the time the planet started warming up again. Although I don't like the idea of living in an ice age, our ancestors had had plenty of time to adapt to this chilly epoch, and they knew how to survive in it. Modern humans had been around for a couple of hundred thousand years at this point, and although the world they were used to was much colder than the one we know, it was also home to many more large animals than we see today. Massive beasts such as mammoths and giant ground sloths roamed the tundra, affording our ancestors terrific hunting opportunities.

After people had adapted to this ice-age environment, the warming planet would have given them some serious headaches. The change in temperature created new opportunities for finding food — along with new challenges. Luckily for both our species, dogs were ideally equipped to help humans solve these new problems.

Humans' excellent vision had made our puny ancestors among the most successful hunters in the ice-age environment of steppes and open pine forests. We had developed effective weapons for operating at a distance: spears, atlatls, and bows and arrows all extended human reach and made us formidable predators. At the end of the last ice age, however, forests that had once been sparsely dotted with trees (think Scandinavia and northern North America) thickened into dense masses, which were far more difficult for humans to navigate, and our powerful sense of sight became useless as thick undergrowth filled in the lower levels of the woods.

Our ancestors in these transitional environments needed a new technology if they were to hunt successfully in this strange new world. This technology had to provide the ability to detect prey through the thick undergrowth of the denser forest understory, as well as the capacity to move quickly through it. This technology also would need the motivation and speed to chase after and corner prey, yet also be able, or at least willing, to abstain from completing a kill on its own. Having found and trapped the target animal, it would need to call out, so the human hunters would know where it was, and then wait there for them to catch up and dispatch the prey. One more thing: this technology would carry minimal risk of harming humans.

Wolves do not possess these qualities — but these skills are well within the capacities of dogs. Dogs inherited a highly sensitive nose from their wolf ancestors, which enables them to find prey under conditions where vision is useless. Dogs also inherited from their forebears the motivation to hunt, and they are typically small enough so that penetrating thick forest is not a challenge to most of them. Yet their ability to complete a kill has faded away to a very considerable degree, leaving them willing to call for help at the final stage of a hunt. Dogs' suitability to this array of tasks provided essential support to our hungry ancestors. As humans struggled to adapt to an unsettling warmer environment, dogs must have seemed like magical beings.

I suspect the partnership of hunter and dog started as an accident: some early dogs, snacking on a village trash pile, decided to follow some men out on a hunt. But I'm sure it quickly grew into a powerful relationship, characterized by strong emotions on both sides. This, I think, is when the connection between human and dog really kicked up a notch into the strong emotional bond we are accustomed to today. Scavenging had created an evolutionary niche for canids that could tolerate humans; hunting would have given these proto-dogs a chance to prove their worth to people. As I'll explain, hunting alongside humans also would have favored exactly the kind of genetic mutations that have made dogs the loving animals they are today.

To really understand how dogs could have helped our ancestors hunt—and how emotion in general, and love specifically, could have played a role in this bonding pursuit—I needed to see for myself what it is like to go hunting with dogs.

I started reading about what anthropologists had discovered concerning different peoples around the globe who still hunt with dogs in something like the way our ancestors might have done. In the process, I came across the work of a researcher at the University of Cincinnati, Jeremy Koster, who had carried out detailed analyses of the hunting practices of the Mayangna people. The Mayangna are an indigenous people living in the Bosawás Biosphere Reserve, in a remote part of Nicaragua by the Honduran border. They practice agriculture, growing beans, plantains, and rice, but in addition, and as Koster's studies make clear, hunting with dogs offers a real benefit to these people. Meat from the hunt is one of the few sources of high-quality protein in their diet.

As luck would have it, shortly after I found his academic papers, I was in Cincinnati at a conference and reached out to Koster, suggesting we grab a beer. Perhaps we grabbed a few too many beers, because the next day I realized I had agreed to travel to Nicaragua with Koster on his next trip to visit the Mayangna.

Koster assured me that it was really easy to get to the Mayangna settlement of Aran Dok, where he was conducting his research. It would take just three days, by road and boat, from the Nicaraguan capital, Managua—itself only a two-and-a-half-hour flight from Miami. What he didn't mention was that the day traveling by road would be spent crammed into the front seat of a Toyota Land Cruiser with two other passengers, over roads that became progressively more potholed and bumpy. The two days by "boat" turned out to be two days in a dugout canoe. A large dugout canoe with a motor, but a dugout canoe nevertheless. It was the most uncomfortable journey I had ever taken.

But once we had passed through the rocks and whitewater into Mayangna territory, the experience was mind-blowing. It was as otherworldly as stepping into Jurassic Park. All that was missing was the dinosaurs, but what we found was almost as stunning: dogs who were living with people in the kind of relationship that their ancestors may have shared with ours many thousands of years ago.

The Mayangna live in sturdy wooden huts on stilts along the riverbanks. As we motored into view, they rushed to the shore and stared, with some anxiety, at these strangers. However, as soon as I waved at them and smiled, they waved back, with big grins and great enthusiasm. People who recognized Koster greeted him extremely warmly. At one point we nearly capsized as we came alongside a smaller dugout canoe containing four men, who each wanted to give Koster a big hug.

After we had strung our hammocks in the guest hut and enjoyed a bowl of rice with a few small hunks of meat for dinner, and another bowl of rice without any meat for breakfast, I set out hunting with some Mayangnan men (only the men hunt). The guys pulled on their rain boots, grabbed their machetes, yelled for the dog, and they were off!

At first, I was struck by the similarities between the Mayangna hunting expedition and the walks in the woods I had taken with my childhood dog, Benji, when I was a kid. First rule: leash your dog. The Mayangna don't have collars and leashes, but they do have rope, which they loop casually round the dog's neck. The leash stays on only while you are making your way through the village. Once you reach the forests, you take off the leash and your dog runs free.

At this point, Benji and the Mayangnan dogs appeared to act in much the same way—but the human behavior was quite different. When I was a kid walking Benji, it was important not to allow him to wander off too far. I would be in trouble if I couldn't bring him home with me. He would get all excited about the scents and sounds he detected in the woods near our home, so I had to call to him to keep him in sight. For the Mayangnan men, by contrast, the whole point of the expedition is for the dog to run off, chasing down whatever it detects in the thick rainforest. If their dog stuck close to them, the men got annoyed and remonstrated with him to get to work. Every now and then they would stop on a hilltop and listen for the dog—occasionally shouting out "Sulu"—the word for "dog" in their language, with the "u" sound very elongated ("Soooo-loooo"). They hoped to hear excited barking or yapping, an indication that the dog had found something. If this occurred, the men charged off as rapidly as they could to catch up with him.

Mayangnan man hunting with his dog

As they raced to catch up with their dog, the Mayangnan men were able to slash their way through the rainforest with their machetes faster than I could keep up, just trudging through the path they had cut. Probably because of the slow gringo they were dragging along with them, we didn't catch anything on either of the hunts I went on, but I nonetheless got a great feel for the process. I could see that it wasn't rocket science — the dog didn't need any particular training. The operation depended on inclinations and capacities intrinsic to the dog: detecting and chasing prey, coupled with an inability to complete the kill on its own. Upon finding and trapping an animal, the dog calls out to the humans — although whether the dog is barking out of frustration, or in the knowledge that its people would come to finish the kill, I cannot say. Either way, the effect is the same: the people come running and complete the hunt.

Going hunting with the Mayangna brought home to me how important it is that hunters' dogs do not kill prey for themselves, but rather cry out and thereby bring the humans to them. If dogs behaved as wolves do, and just killed and ate what they found for themselves, they would be of no assistance to people. This underscores the impossibility that our ancestors could have hunted with wolves. Rather, they had to wait for dogs to come into existence before they could involve such a useful companion in the hunt.

It is a testament to the power and endurance of the human-canine bond that these furry little hunting companions are equally effective in the present day. According to Koster's data, the dogs of the Mayangna typically weigh about twenty pounds, yet each dog brings home, on average, over ten pounds of meat each month. This is an impressive contribution to the protein needs of the people. Because of this, a successful outcome leads to a great outpouring of emotion, shared by the people and their dogs. This positive experience doubtless strengthens the bond between the people and their canine companions.

In Aran Dok, the main village of the Mayangna, two of the men own rifles. Koster found that the scrawny dogs were just as effective at bringing in prey, on average, as the guns.

Watching the Mayangnan men and their dogs hunting together, I was also struck by the intensity of the bond between them. I could see that for dogs, helping people hunt required a whole different set of skills than scavenging did. Picking over trash dumps is a pretty solitary pursuit; dogs busy digging in the town dump are not interested in company—human or canine. On the other hand, when I was out in the rainforest with the Mayangnan men, I got the strong impression that this activity demanded coordination and mutual understanding between dogs and men. Its success depended on accurate communication. The men let the dogs know that it was time to look for prey, and the dogs should detect and chase down some. For their part, the dogs, once they found prey, had to communicate this to the people, indicating where they were in the dense forest. The hunters even claimed that, from the tone of the dog's cry, they could tell what the dog had caught — but since we caught nothing on the two hunts I went on, I wasn't able to verify this for myself.

Upon returning from Nicaragua, I became slightly obsessed with the question of how hunting might hold the key to explaining why dogs developed the ability to love humans. As an acolyte of Ray Coppinger, I had been hesitant to think that hunting had played a role of any significance in the origin of dogs and their capacity for interspecies relationships. Not only had Ray poked holes in the idea that humans had "created" dogs to serve as hunting companions; he also had been skeptical about the possibility that people long ago had found much benefit in hunting with dogs. He suspected that it took too much effort to train a dog. He thought the whole hunting thing was more of a "male display," which men had used to impress the ladies, rather than a practice with economic benefits.

But now I was reconsidering my position. Even if it hadn't started dogs down the evolutionary path that eventually would distinguish them from wolves, I wondered whether hunting did help canines go the extra mile.

Dogs with genetic mutations that made them more inclined to form strong bonds with people, I surmised, would have been at an advantage over those who remained aloof. These friendlier dogs would have been more likely to follow people out on a hunting expedition and to call for human assistance in completing the kill, and thus would have had a better chance of sharing in the profit of the hunt. This would have led to better odds of survival and more pups, which in turn would mean that these friendly dogs' genes eventually would have become omnipresent in the tribe.

I wondered whether my archeologist friends could point me to evidence that would shed light on the possibility of a powerful relationship between people and dogs back when the ice age was ending and our ancestors needed help in the hunt. Angela Perri, a zooarcheologist at Durham University in the UK who has a special interest in the importance of dogs to our ancestors, was happy to oblige. She showed me that there is indeed proof that, around the time that hunting with hounds started to catch on, our ancestors left signs that they cared very deeply about their dogs — evidence that a strong emotional relationship between dogs and people was developing in correlation with their burgeoning predatory partnership. While correlation does not prove causation, of course, Angela's research nevertheless points to a strong connection between these two milestones: humans and dogs hunting together, and the formation of strong emotional bonds between our two species.

For her PhD research, Perri focused not on the burials of people together with dogs, but instead on careful burials of dogs alone. Her reason for this emphasis was that there are many reasons why an animal might be buried with a person — most of which needn't tell us anything about a possible relationship between the deceased person and the dead beast. The room at the Israel Museum in Jerusalem that contains a resin copy of the bones of the woman buried with a puppy twelve thousand years ago also has vitrines containing people buried with deer antlers, with tortoiseshells, with fox teeth, and with a variety of other animal parts. None of these should be taken to imply that the people of those times were developing emotional relationships with deer, tortoises, foxes, or whatever. The people who buried this woman had their own now-lost ritual reasons for putting bits of animals in graves with their relatives.

When you think more deeply about the burial of a dog with a deceased person, you can't help but wonder how the dog ended up there. Did it die fortuitously around the same time, or was it killed intentionally to decorate the grave or perhaps accompany the deceased person on the voyage into the afterlife? Given that a pet's spontaneous death from natural causes at around the same time that its master or mistress dies cannot occur all that often (though Charles Darwin's last dog, Polly, died three days after her master breathed his last), the majority of co-burials of dogs with people must represent intentional killings of the dogs. Of course, we have no way of knowing what the people of many thousands of years ago were thinking. It is not perhaps entirely impossible that they had loving relationships with dogs, yet this did not preclude the possibility of killing a dog in order to bury it with a person that the dog apparently loved.

As Perri points out, the emotional implications of dog-human co-burials are ambiguous at best. But much clearer inferences can be drawn from cases where people bury dogs on their own, with great care and respect.

When there is no human in the grave, the dog's meaning to the people who buried it is unambiguous. If, as our ancestors did at certain periods, a dog was buried with as much care in a grave as richly decorated as any human interment at that time, then we have a clear signal of how much those people cared about the dog.

Perri analyzed ancient dog burials in three parts of the world: eastern Japan, northern Europe (including Scandinavia), and a region of the eastern United States that covers parts of Kentucky, Tennessee, Alabama, and bits of other states. She reviewed reports on hundreds of dog burials in those three diverse parts of the world. She looked at when the dogs were interred, and how. Were they buried with rich grave goods and other signs of care and respect, or were the internments casual and seemingly incidental? In other words, were there signs of mutual love and affection, or were people just getting stinky old dog carcasses out of their way?

What's so interesting about the widely dispersed regions that Perri focused on is that crucial developments in human history took place at quite different times in the three locations. The end of the last ice age, the difficulty our ancestors had with hunting in ever-denser forests, the induction of dogs as hunters' helpers, and finally the development of agriculture, which lessened humans' reliance on hunting, occurred at times that diverged by thousands of years in these three places.

Perri made an amazing discovery. For each location, she plotted a graph with the number of careful, intentional burials of dogs across time — from back in the ice age until relatively recent times (recent for an archeologist still means some thousands of years ago). In each case she found the graph took the same general form — a simple inverted-U shape. If we go far enough back in time at each location, people did not bother burying their dogs with any particular conscientiousness — a low point on the graph. Move far forward in time and the graph also dips low: people didn't bother much then, either. But in each location there was a central "hump" in the graph — a lengthy period of time when people in each of the three parts of the world put enormous care and effort into burying their canine companions.

The exact dates of this period differed in each location, but the point in human history when it occurred was always the same. People maximized the care they put into burying their dogs during the period following the end of the last ice age, when the planet was warming, and hunting was getting more difficult. During that phase of human history, and after hundreds of thousands of years of hunting very successfully on their own, our ancestors found themselves stymied by woods they just could not see or move through. It is in that window of time that people took to burying their dogs with great care. And these peoples were scattered at three widely separated locations on the globe. During this period, any time from three to nine thousand years ago at the different sites (more recent in northern Europe, earlier in North America), these people could not possibly have known anything about one another. They must have made their decisions to treat dogs in this new, caring way completely independently. In each location, these practices tapered off with the advent of agriculture.

In some cases, the dogs that Perri analyzed were buried with such riches that the archeologists who originally discovered them could not believe the remains could really be just dogs. One archeologist proposed that the dogs were "cenotaphs" — animal bodies buried as substitutes for lost human warriors. I think Perri makes a great counterargument. These ancient people understood these dogs to be dogs—they were not stand-ins for valued humans—and buried them with lavish grave goods because the animals had proved their value by helping in the crucial activity of hunting. Quite possibly, our ancestors also interred these dogs with high honors because they had expressed strong affection toward the people around them, so that the people felt compelled to reciprocate.

Taken together, this archeological evidence strongly suggests that, although helping human hunters likely isn't what created dogs, being indispensable tools for capturing protein did spark the powerful bond of affection between human and dog. What the archeological record doesn't tell us—at least not yet—is whether the genetic mutations that allowed dogs to reciprocate their humans' love also occurred during this precise period.

Our paleo-geneticist friends are not yet finished with the genetic analysis of ancient dog bones, but with a little luck these remains will tell us when the genes for hypersociability—the genetic basis for dog love—first began to appear in the canine kind. Short of an interview with a human ancestor of seven, eight, or nine thousand years ago, this genetic analysis is the next best report we can have of how the dogs of long ago interacted with people.

Though I look forward to having that genetic evidence in hand, I still can't help but be sad that we will never have a blow-by-blow account of how, where, and when dogs became the loving beasts that they are today. Of course, the people who were on hand during this critical phase of dogs' history are long gone, so I'll never get to do that interview. I've done my best to content myself with the research at hand, which, happily, goes beyond even the findings from the examples I've discussed so far.

Today we have scientific evidence of how dogs might have emerged, in a relatively short time span, out of the genetic tapestry of their wild ancestors. The evidence comes to us not from wolves, but from another close canid relative: the fox. And it comes not from the chilly landscape of ice-age Europe, but from Soviet-era Siberia, of all places. Starting in 1959, one of the largest-ever experiments related to evolution was carried out: a direct experimental test of whether evolution can create love.

Soviet-era Siberia might seem like an improbable testing ground for the history of dogs' love. Whereas the early Soviet Union had pioneered genetics, Stalin disapproved of this bourgeois science and by the 1930s was having geneticists sent to the gulags and even murdered.

With Stalin's death in 1953, however, came a resurgence of genetic research in the Soviet Union. One of the leaders of the new generation of genetic scientists was Dmitri Belyaev. His brother Nikolai, also a geneticist, had been executed for his scientific beliefs in 1937.

Dmitri Belyaev wanted to vindicate his murdered brother in an experiment that would demonstrate that evolution doesn't inevitably lead to a stark conclusion about nature — that it is "red in tooth and claw." By contrast, it can form a pathway to affection — even love. Belyaev wanted to prove that friendliness toward humans could be inherited — a radical concept at the time. Although it was known that bodily forms were heritable, it was far less clear that complex patterns of behavior could be subject to evolution.

To investigate these issues, Belyaev chose to work with foxes. Fox furs were very important in the frigid Soviet Union, but foxes were also a smart choice for an investigation of what underlies the origins of dogs' remarkably affectionate nature. Like dogs and wolves, foxes are members of the family Canidae, but unlike dogs and wolves, they are not part of the genus *Canis*. This is significant. It means that foxes are sufficiently closely related to wolves and dogs for an experiment on foxes to shed light on the origins of dogs. But foxes are also distinct from wolves and dogs, enough so that we can be confident that foxes have never interbred with either of them. This means that whatever Belyaev might uncover in his experiment could not possibly be contaminated by interbreeding of the foxes with dogs or wolves.

Each spring, Belyaev selected those foxes that were least fearful and most friendly toward humans to be the parents of the next generation. In the third year of his study, already a few of the foxes were letting themselves be held, rather than acting with the meanness characteristic of wild foxes shut up in cages. Ember, a member of just the fourth generation in the experiment, was the first fox ever to wag its tail in excitement at seeing a human approach. By the time of Belyaev's death, in 1985, it was abundantly clear that his experiment had been a complete success.

When I first heard that the Soviets had attempted an experiment on the evolution of love in Siberia in the 1950s, it just seemed so crazy, I couldn't believe it. Now that I have visited the fox farm of the (ex-) Soviet Academy of Cytology and Genetics for myself and become as well read about what went on there as I can, I understand that what the Russians did was nothing less than what they claimed all along. It might not be the stuff of a James Bond film, but the truth is far more astonishing than the plot of any hackneyed cold-war melodrama, some residual adolescent nostalgia for Honey Rider and Pussy Galore notwithstanding.

Growing up in Britain during the cold war, I had been taught that the USSR was an evil empire bent on taking over the planet. But I had never really comprehended just what an enormous landmass Russia occupies until I got on a plane to Moscow, and then flew for another three hours from Moscow to the largest city in Siberia, Novosibirsk, an industrial behemoth located where the Trans-Siberian Railway crosses the River Ob. On the map, Novosibirsk is not even halfway across Siberia, and yet it felt very different from Moscow, and a totally other planet from the Florida I had left the previous day.

The trip from the Novosibirsk airport out to the Laboratory of Evolutionary Genetics of Animals (known to all simply as the "Fox Farm") took me past factories so decayed, only clouds of dark smoke from tall chimneys revealed that they were still in operation; past little babushkas bundled up against the already cold September weather and sitting on upturned buckets, selling produce and flowers; past a memorial to a collective farm; and past potholes so enormous, they looked more like bomb craters. Finally, after about half an hour's drive, we reached the entrance to the research station.

Inside the gates of the farm, there was derelict fox caging everywhere and collapsed or collapsing concrete buildings in many places. Weeds and grass had reclaimed much of the compound from human use, but local people had also laid claim to the land. I saw a man harvesting potatoes in one area, and flowers grew between some rows of fox cages, beautifying such an unpromising location.

We walked slowly around the old cage structures, looking at the animals. The tame foxes whimpered and trembled in excitement at our arrival, seemingly desperate for human contact. They reminded me of dog pups, in their engaging, extroverted enthusiasm for people. One of my guides opened a cage and a fox literally jumped into her arms — a truly astonishing sight. The fox was passed over to me and seemed very excited to be held by me too.

I may have been a newcomer to fox cuddling, but this animal was determined to teach me. It gave little whimper-squeaks, wagged its big fluffy tail, and nuzzled into my neck. Several foxes were removed from their cages and passed around: they all showed the same re-

The author with one of the descendants of Belyaev's friendly foxes

action. At first they shook in excitement, but then quickly calmed down and seemed to greatly enjoy being held. I had my photo taken with tame foxes of different colors. Each put its face close to mine in the most intimate way. They may look like foxes, but in a deep sense Belyaev had created a new beast, much more like a dog.

What I saw in Siberia shows how loving domesticated animals can be created out of wild ones that have no interest in interspecies relationships. It shows that, as Dmitri's murdered brother Nikolai believed, selection can be an enormously powerful force in dramatically changing animals in just a few generations. Belyaev's long-lived fox experiment teaches us that it is possible, by selection alone, to create an animal as tame as any dog. It lifts the veil, at least a little, on how evolution might have given us our best friend in the animal kingdom. It is the nearest thing we will ever have to a direct experimental demonstration of how friendliness and affection was bred into dogs.

As important as it is to identify what Dmitri Belyaev was able to demonstrate with his experiment on foxes, it is just as important to clarify what his experiment does *not* prove. Unfortunately, Belyaev's team did not actually *do* anything with their foxes, beyond feed them and breed them. They did not go out hunting with them, or attempt any other kinds of cooperative tasks. Thus, we cannot deduce directly from this experiment that any particular activity, like hunting, led dogs to develop their friendly nature. That will just have to remain conjecture, backed up by archeology and anthropology.

Further, it would be a mistake to assume that, because Belyaev and his coworkers in Siberia hand-picked those animals that would beget the next generation of foxes at the farm, ancient humans must have done the same, selecting which animals would become parents in the history of the dog. Selection is selection, whether it is carried out by people or by nature. As Darwin himself pointed out, "artificial selection" (human choice of who shall become the parents of the next generation) is just a pale reflection of "natural selection" — the struggle to leave a biological legacy, which takes place in nature without human intervention. Both can lead to the same results. Belyaev's epic experiment shows that selection can produce friendlier animals. It does not tell us who — or what — carried out that selection in the case of the dog.

As I already said, I do not believe that humans created the first dogs. While people the world over breed dogs today, I cannot imagine that our ancestors controlled the mating of animals. Back when dogs were first coming into existence, humans had none of the technology of collars, leashes, cages, or even walls and high fences that are essential to control the sex lives of another species.

It is not inconceivable that our ancestors could have carried out the practice biologists euphemistically call "post-zygotic selection": that is to say, culling pups they didn't like the look of. But even that was probably too hit-or-miss to have much impact. If you think you are going to cull some of a wolf mother's pups, I say "Good luck." My guess is that it would be easier to cull all the pups, and the mother as well, than to get at some of them and leave others behind. Only by selectively culling some and leaving others to become the parents of the next generation could you hope to make any progress in turning wolves friendlier — and frankly, I just don't think it could be done.

Our ancestors also likely lacked the understanding of inheritance that is essential to human intervention in the breeding of other species. It is only in highly inbred animals, such as purebred dogs, that traits "breed true" anyway. If you have two white purebred dogs, it is highly likely that their offspring will have white fur too. But if you have two white mongrels, their pups may come out all sorts of different colors. Genetics is a very complex subject, one I don't fully understand today, and I don't believe my ancestors over fourteen thousand years ago had the foggiest idea of how traits were inherited.

On balance, I am convinced that it must have been natural selection that created the dog. The advantages of tolerating people would have been so great for the population of wolfy animals who had made their homes on our trash dumps that they would have been selected by nature for the capacity to at least allow people to get closer to them. When the ice age ended and our ancestors needed help with the hunt, dogs' tolerance for humans gave way to the open, loving affection in whose glow we still bask today.

One thing is for sure: dogs were created by genetic changes occurring over the course of generations. Just how many generations it took for dogs to become the animals they are today is something we may never know. It is possible that a random mutation or two caused dogs to transmute quite suddenly from creatures that simply put up with people to the endearing, engaging beasts we know and love today: animals who do not just tolerate us, but positively seek us out and convince us to care for them. These animals can convey to you, when you are at the shelter, looking for a new canine companion, that *they* have chosen *you,* rather than the other way around. Exactly how this creature's genome differentiated itself from that of the wolf is the subject of some of the most interesting research in canine science today.

But no dog is the product of his or her genes alone. Rather, every individual dog's quirks — loving behavior included — are the outcome of a delicate interplay between its genes and its environment. How a dog's life makes it a loving being is a fascinating subject in and of itself. And for those of us who share our lives with dogs, the question of how affection can be nurtured in these precious animals is perhaps even more important than the question of how they were primed for love in the first place.

6

HOW DOGS FALL IN LOVE

THE GENES OUR DOGS possess are key to what makes them special. But these genes don't determine the form of the finished product the way that the instructions for, say, a set of Legos guarantee the shape of the finished toy (assuming you haven't lost any bricks). Each organism's genetic blueprint is, rather, the starting point in a developmental process that would create a hundred different organisms if it were repeated one hundred different times.

My own cherished, lovably idiosyncratic dog reminds me of this constantly, albeit unintentionally. As Xephos rests behind me right now, she has half an ear and half an eye open for anybody who might make a delivery to the house, and an awareness of what I'm up to, hoping for the (unfortunately rather slim) chance I might get up and take her out for a walk or a car ride. To be able to do this she must, of course, have genes that code for the proteins to make eyes, ears, and a brain capable of making this kind of information processing possible — but it takes more than genes for Xephos to actually do what she does. To be who she is, with her tremendously sweet disposition and her particular likes and dislikes — that takes not just genes, but also a particular set of life experiences.

Obviously, genes play a role in the story of dogs' love, just as they have a role in every story in biology. And the discoveries of genetic differences between dogs and their wild ancestors, especially the genes that contribute to our companions' warmhearted nature, are among the most exciting developments in dog science of recent years. But the world around a dog shares responsibility for who she becomes.

Even with all the right genes for love, a particular individual dog is not guaranteed to become a being who loves people. That takes nurture, as well as nature. To reintroduce terms that served as lodestars in my journey of discovery, the story of dogs' love is not just a tale of phylogeny (evolutionary change over generations), but also one of ontogeny (an individual's personal development). Which, of course, invites the million-dollar question: if dogs are empowered by evolution to love humans, but are not required to do so, how do they come to love us at all?

I was thinking about the contingencies of dogs' love when I stumbled across some articles in the popular press about how the singer Barbra Streisand had had her much-loved Coton de Tuléar, Sammie, cloned. Cloned animals share all of the genes of the animal they are derived from; in this way they are, like identical twins, genetically completely indistinguishable. If you want to compare the effects of phylogeny and ontogeny, I knew, you couldn't do much better than identical twins. Scientists have been studying them for decades to learn more about how human beings are shaped by the complex interplay between our genetics and our environment. Could clones, I wondered, offer a similar insight into the nature-versus-nurture question as it relates to dogs?

Most of the articles I read about Streisand's dog Sammie focused on the enormous expense and ethical concerns related to the practice of cloning. The first dog was cloned in 2005 in Korea, in a process that involved the implantation of eggs into 123 surrogate mothers to produce a single viable offspring. The ethical implications are clearly highly problematic when so many female dogs are being used in this way. In the intervening decade or so, the process has become streamlined, and an organization in Texas will clone your pet for you using a single surrogate mother, if you give them some cells from inside your dog's cheek. That, and $50,000.

I certainly share the widespread astonishment at the amount of money involved in pet cloning, and the ethical issues trouble me too — but what I found most interesting was what Ms. Streisand herself said about the dogs. She reported that the four pups produced by the cloning process looked the same, but, she also wrote in the *New York Times,* "Each puppy is unique and has her own personality. You can clone the look of a dog, but you can't clone the soul." I thought this was a really intriguing remark. What did she mean exactly by saying "You can't clone the soul"?

Unfortunately, Ms. Streisand did not respond to my attempts to make contact, but I did discover a man living only twenty minutes away from me who had his dog cloned in 2017. Like Barbra Streisand, Rich Hazelwood sent $50,000 and some cells from the mouth of his beloved terrier-cross, Jackie-O, to Texas; five months later, he had two new dogs, which he named Jinnie and Jellie. When I spoke to him by phone, Hazelwood told me that, although the clones look fairly similar, their personalities are "as different from each other as they can be. Jinnie is totally her mother. She is a total athlete — a hunter, a runner. She can run three or four miles without stopping." Although Jellie had exactly the same DNA as Jinnie, she could not have been more different. "Jellie is kinda a couch potato," Hazelwood said. "Very smart but not very active."

The clones, Hazelwood told me, also differ noticeably from their mother (or sister, or DNA donor — or whatever term you prefer). Jackie-O is three-quarters Jack Russell terrier, with the remaining quarter made up of black Scottish terrier and English bulldog. It's a beautiful mix, producing a small dog with short, curly hair — mainly white, with some brown patches. Jinnie and Jellie have facial markings similar (but not quite identical) to those of Jackie-O. But whereas she has a brown patch across her rump, Jinnie and Jellie are entirely white from the neck on down, a testament to how even tiny differences in an animal's early life, including life in the womb, can influence the precise form its body takes.

Hazelwood's two cloned dogs certainly look similar enough to be twins, but when I visited them with Lisa Gunter, then my graduate student and currently my colleague and collaborator, their behavior was, as he had said on the phone, completely different. Jinnie ran up to us and around us, jumped up to greet us, bounced on our laps as soon as we sat down, and was on the alert the whole time we were there. Jellie also came over and greeted us, but soon she was napping on the sofa.

Amazingly, the two young clones' mother (or sister . . .) is still alive. Jackie-O, now eighteen years old, came over to us — and barked. And barked. The poor old lady is now blind, and I suspect pretty deaf. She was friendly and surprisingly mobile for a dog of her age, but not able to keep up with her daughters. She stayed on the floor, and it took quite a while for her yapping to stop. Hazelwood said that, when she was younger, Jackie-O had had a very exuberant personality, as Jinnie now did.

Rich Hazelwood's genetically identical — but temperamentally distinct — cloned dogs, Jinnie and Jellie

Before I visited with Hazelwood and his trio of genetically identical dogs, I had felt cynical about the value of dog cloning, and I still wouldn't encourage the practice. But it was hard to maintain my skeptical position in the face of Hazelwood's intense joy at being with these dogs. He explained that a few years back he had been at a low point in his life, and the thought that his beloved Jackie-O could not be with him much longer had burdened him. When he heard that it was possible to clone dogs, he was on the phone to Texas in a flash.

This is how he summed up the outcome: "The joy I experience in my life from this experience is worth every shekel of the fifty grand." It was very hard, as Jinnie sat on his knee and Jellie slept next to him on the sofa, to begrudge Hazelwood the obvious pleasure he took in the two dogs.

Seeing clones with my own eyes was quite startling. I knew from basic scientific principles that personality could not be fixed entirely by genetics. But I still imagined that two individuals who had the exact same genes, had been carried at the same time by the same mother and born within an instant of each other, had been reared in the exact same environment, and had gone on to live in the same home, would have similar behaviors. Streisand's comment about the impossibility of cloning souls had made me think about the range of variation of personality that might still be possible, even with so much held constant. Yet I was still stunned by the striking differences between Jinnie and Jellie. We did not attempt any formal tests of Hazelwood's dogs' personalities, but I would put Jinnie in the top 20 percent of the most extroverted dogs I have known, whereas Jellie belongs among the more introverted ones.

Clearly, the tiniest differences in a dog's life experiences can have an enormous impact on the way that their genes are expressed. To put it another way, a dog's DNA is not its destiny. And this principle holds whether we're talking about a whole slew of genes, such as those shared by Jinnie and Jellie, or a narrower suite of genes, such as those that give dogs the ability to express love.

Puppies are born with the genes for love, but it still takes a village to raise a loving dog.

When we look at dogs' loving nature, they certainly could not be who they are without genes that make affection toward members of other species an option. But the right upbringing is just as essential for this pattern of behavior to express itself. It is widely understood that pups can be raised to be aloof and even aggressive toward people if their upbringing encourages those patterns of behavior. Yet few people are aware that young pups can grow up loving species other than just our own, and recognizing this fact is tremendously important for understanding how it is that dogs grow up loving us.

I'm going to let you in on a secret here, and I hope it isn't too upsetting. Dogs love us, yes, but their love for us isn't about us. It's about them. Your dog loves you, but your dog could love almost anyone —and not just any human being, but any being, period. Had your dog been reared by aardvarks or zebras, it would have grown to love them just as it now loves you.

The capacity that dogs have for loving people is just a capacity for loving—it isn't specifically focused on our species. This seems surprising to you only because you are a human being, and so you mainly see dogs interacting with others like you. It's a forgivable mistake. It might be an exaggeration to say that a dog reared with aardvarks will grow up loving that species—I don't suppose it has ever been tried—but I do know for a fact that, if you had been brought up as a goat being protected by a livestock-guarding dog such as an Akbash or an Anatolian shepherd, you could be forgiven for thinking that dogs loved only goats.

Even penguins have been the beneficiaries of dogs' love, on an island outside the small community of Warrnambool, about two hours' drive along the Great Coast Road west from Melbourne, Australia. Just off the shore, the not exactly poetically named Middle Island provides a home to a community of little penguins. Little penguins are not just small penguins; they are a distinct species of penguin,

Eudyptula minor, native only to Australia and New Zealand. I have visited with little penguins on the (also rather prosaically named) Penguin Island in Western Australia. They are surely the sweetest members of a group of birds already high on the cuteness scale. Little penguins typically stand about one foot tall, and their back is a shade of blue between slate and navy. Where their larger cousins, such as the emperor and king penguins, can appear businesslike, perhaps even a little severe, little penguins, because of their size and cheerful waddle, come across as sweeter and even playful.

To see the little penguins on Penguin Island, I walked across a causeway at low tide. The local government discourages walkers because there is some risk involved if the weather changes suddenly, but I didn't come to any harm. Being a little under half a mile long and always covered by at least a few feet of water, the causeway does a good job of keeping the penguins safe from predators on the mainland.

Unfortunately, the penguins out on Middle Island are not so lucky. Their island is just sixty feet offshore, and shifting sands on the beach can create conditions enabling almost anything to walk across. Tragically, in 2004, foxes marched onto the island and killed almost all the birds. The penguin population of the island had once numbered over eight hundred, but in 2005 only six birds were found. The local townspeople were distraught, but they had no idea of what to do. They could hardly move the island farther offshore.

A nearby chicken farmer, known as Swampy Marsh, had the brilliant idea of deploying dogs raised to protect livestock to guard the penguins. Marsh had despaired of shielding his free-range chickens from foxes until he acquired a Maremma dog to guard them. He was impressed by the Maremma's prowess at frightening foxes off the property. Marsh's first Maremma, Ben, chased a fox out into the road. Marsh told the *New York Times,* "It got squashed. It was fox pizza."

Maremmas are an ancient breed of dog from a part of southern Tuscany that a British newspaper termed "chic but discreet." Other livestock-guarding dog breeds are found all around the northern Mediterranean, from Portugal to Turkey and into the Middle East. The use of dogs to guard livestock was practiced in southern Europe thousands of years before tourism took over as the major moneymaker in that part of the world. Homer, in the *Odyssey*, a tale reckoned to be over three thousand years old, reports that Odysseus, on his return to Ithaca, was nearly killed by dogs guarding pigs.

Livestock-guarding dogs are not the sheepdogs of old England. Sheepdogs that herd are a quite different kind of beast. Their task is to follow closely the instructions from their master in order to corral livestock — usually sheep — moving from one place to another. Although I'm sure I will attract hate mail from advocates of breeds like the Old English sheepdog, I doubt that herding dogs are nearly as ancient as livestock-guarding dogs. For one thing, guarding livestock is much simpler than herding it — it does not involve elaborate direct instruction, by shout or whistle, from a human master. For another, the historical evidence for dogs carrying out guarding functions goes back to the beginnings of recorded history — the evidence for herding appears later. The first sheepdog trials were a late-nineteenth-century development.

Unfortunately, the concept of using dogs to guard livestock was something that Europeans failed to bring with them when they populated North America. It wasn't until the 1970s that it was reintroduced, largely through the efforts of Ray Coppinger, at Hampshire College. Ray took many trips to southern Europe, visiting isolated mountainous regions of Portugal, Spain, Italy, Greece, and Turkey. There, he watched the shepherds and learned how they used dogs to guard their flocks. He brought back to North America the first Maremmas the continent had seen.

One day, the chicken farmer Swampy Marsh was chatting with a biology student named Dave Williams, who was working on his farm, about the tragic situation on Middle Island. Marsh remarked that they ought to put a dog out there with the penguins. Williams made that idea the focus of part of the coursework for his degree. The report he wrote ended up as a proposal submitted to the Warrnambool city council, who—with few other alternatives available—permitted Williams to camp on the island with Marsh's dog Oddball.

Though Oddball became the star of an eponymous movie, her residence on Middle Island was not a complete success. After a week of camping on the island with her, Williams left Oddball on her own with the penguins. The poor dog got lonely, and after three weeks she ran away and went home to Marsh and the chickens. The problem was, Oddball loved people too much and wasn't satisfied with just the penguins for company.

The second dog that Williams and Marsh tried on the island—Missy—stayed a little longer (part of the reason they chose Missy was because she had a bad hind leg, which made it more difficult for her to climb down the cliff and get away), but after a few weeks she too ran home to civilization.

Although the first two dogs had failed to establish a binding relationship with the penguins, they had done enough. That first penguin-breeding season with Maremmas on the island, not a single little penguin chick was lost to foxes.

Williams knew how to improve the situation. For livestock-guarding dogs to be motivated to truly care about their charges, they need to be exposed to them early in life. Today the little penguins of Middle Island are guarded by two dogs, Eudy and Tula (from *Eudyptula,* the Latin name for the genus that the penguins belong to), who have known penguins since they were pups.

Just as dogs have to be exposed to penguins early in life in order to love them, so too do dogs have to meet with people early in life in order to love us. Raised on a farm, pups can form strong bonds with any and all of the animals they interact with there as they grow up: pigs, goats, cows, ducks, chickens, and whatever else the farm may have. If the farm in question has no humans (I believe George Orwell wrote about one such place), then dogs will grow up feeling no love for humans.

Farmers and their ilk have known this quirk of dogs long before scientists came around to it. Back in the 1830s, in Uruguay, Charles Darwin stumbled across dogs guarding sheep without backup from a human shepherd. He enquired at the nearby estancia as to how "so firm a friendship had been established" between the dogs and the sheep. The locals told him that "the method of education consists in separating the puppy, while very young, from the bitch, and in accustoming it to its future companions . . . A nest of wool is made for it in the sheep-pen; at no time is it allowed to associate with the other dogs, or with the children of the family. From this education it has no wish to leave the flock, and just as another dog will defend its master, man, so will these the sheep."

Darwin's findings did not permeate the public consciousness as fast or as far as they ought to have. Indeed, late in life, Ray Coppinger lamented that people had missed the point of how livestock-guarding dogs work. Because he had gone to Europe and brought back breeds such as the Maremma from Italy, the Akbash and Anatolian from Turkey, and others, people got the idea that livestock-guarding breeds had an instinct to guard farm animals — and the rest of the world's dogs would not be inclined to do this. Ray conceded that there probably was a genetic aspect to being a good livestock-guarding dog — such as having only very modest hunting instincts, for example — but he was adamant that early life experiences were the absolutely critical factor in creating a dog that would love and care for livestock.

Ray understood, as did Darwin, that what makes a dog inclined to guard livestock is the experience of being raised with the species it is going to be asked to care for. Opinions differ as to how wise it is to completely isolate the pup from humans and its own kind — it is probably smarter to let the dog have some acquaintance with people, so that it can be handled by humans later in life. Also, the inability to interact with its own species can be a problem for an adult dog's sexual urges — but the centrality of putting the young pup with the species it is to defend is unimpeachable. No amount of selecting for the right genes can make up for the wrong experiences early in life.

What is happening here is a process called imprinting. Imprinting was discovered by the Austrian ethologist (and dog lover) Konrad Lorenz in the 1930s. It is the crucial link between the genes for love—which create dogs' potential for strong bonds with us, but cannot on their own make any individual dog a human-loving creature—and a dog that actually loves people.

Lorenz famously demonstrated imprinting in geese. He arranged matters so that the first thing a group of goslings saw when they hatched from the egg was Lorenz himself, and this experience was enough to ensure that they imprinted on him. There are lots of cute photographs of young geese following Lorenz around.

Imprinting is the process by which a young animal learns who it is. No being is born knowing what species it belongs to and what kinds of things it should form relationships with. Every individual animal must learn the answer to one of life's most pressing questions—who are my kind?—by looking, smelling, and listening, once its senses open up to the world. Young animals of all kinds cast around early in life, and whatever beings they come across they recognize henceforth as the right types of beast to seek out for company through the rest of their lives.

Most animals have only a brief window of time during which they are open to learning whom they should befriend (what biologists call the "critical period for social imprinting"). For wild animals, like the wolves from whom our dogs are descended, it is essential that this business be dealt with quickly. No formal experiment has ever been attempted on wolves, but there are good reasons to believe that the window of opportunity to convince a wolf that it might like to be friends with people is closed off by three weeks of age.

This makes good sense: in nature—*The Jungle Book* and a thousand other children's tales notwithstanding—it is unwise for the beasts in the field and forest to make friends with members of other species. A prey animal that tried to befriend a predator species would quickly become dinner, and a predator animal that made friends with a prey species would soon starve to death. The shortness of the window of time during which a wild animal is willing and able to learn what kinds of beings to form relationships with pretty much guarantees that, barring exceptional circumstances, wild animals make friends only with others of their own kind.

The brevity of a wolf's "critical period" is why it is so difficult to raise these animals to accept human beings as social companions. It is also why Monique Udell and I were so fortunate to be contacted by Wolf Park — one of the world's leading locations for the careful hand-rearing of wolf pups — and invited to test their docile charges. Wolf Park started hand-rearing wolves in 1974. The founder of the park, Erich Klinghammer, had studied at the University of Chicago with Eckhard Hess, who quite literally wrote the book on imprinting. A scientific understanding of imprinting was essential to Wolf Park's success, and even so, the first attempts Klinghammer and his students and volunteers made at hand-rearing wolf pups to accept human company were beset with problems. Some of the old-timers at the park still have scars to show for their early, stumbling attempts to convince wolves that they should make friends with people. Through trial and bloody error, Klinghammer's crew gradually became aware that they really had only a couple of weeks when they could induce the wolf pups to accept them, and during that time they had to be with the pups twenty-four/seven if the relationship was going to stick. In time, their work paid off handsomely — not least in setting the stage for research that is still being carried out today.

In 2010 one of my students, Nathan Hall (now a professor at Texas Tech University), and Ray Coppinger's last student, Kathryn Lord (now a researcher at the Broad Institute), raised a litter of Wolf Park pups in order to study their behavioral development in detail. I got to see the procedure up close. The pups were taken from their mother at ten days of age and brought into the special pup room at the park. This provided just enough space for a foam mattress on the floor and a similar-sized space next to it. Kathryn and Nathan were barely acquainted when they started swapping twelve-hour shifts here: bottle-feeding the pups, wiping up the mess that came out of their nether ends, and trying to catch quick naps while the pups slept, before being woken by hungry mouths and starting the cycle over again. With six pups, it was a lot of work, and whenever I popped up to see them, Kathryn and Nathan were always very bleary-eyed. And yet the labor was clearly tremendously rewarding, and the affection they developed toward the pups was visibly reciprocated. Because these young animals were still in the critical period for social imprinting, they were ready to love Kathryn and Nathan and learn to accept people in general as social companions.*

In the course of this roughly seven-week ordeal, Kathryn and Nathan learned something that lion tamers have known for centuries: taming wild animals is possible, but it is hard work. While wolves and lions alike can be socially imprinted onto human beings, the window of opportunity when that can be done is very short, and for the relationship to stick, the exposure to people must be maximal.

Taming dogs, on the other hand, is so simple that many people do not even realize they are doing it. Just like wolves and lions, each individual dog pup has to imprint onto people at an early age if it is to accept humans as social companions throughout its life. But unlike these other members of the order Carnivora, dogs are easy to tame. If a dog pup is born and grows up anywhere near people, it will form enough of a social bond with humans to be friendly with them for the rest of its life. Even street dogs that do not reside within a person's home nonetheless usually grow up to treat people as companions if they come near enough that the pups can hear, see, and smell people sometime during the first few months of life. Such is

---

* Around eight weeks of age, the pups were reintroduced to adult wolves. Wolf Park had learned over the years the crucial importance of ensuring that the young wolves become not only socialized to people but to their own kind too.

the power of dogs' predisposition to form strong emotional bonds —to love and seek love.

The proof that dogs' love for people must be nurtured at an early age comes from one of the few large-scale experiments carried out that focused specifically on dogs' behavior. Dogs have never mattered much to the agencies that put up the money for Big Science, so large-scale experiments on dogs are few and far between. Nonetheless, back in the 1950s a major series of studies at the Jackson Laboratory in Bar Harbor, Maine, continued for thirteen years and involved hundreds of dogs. The findings it produced mark it as one of the seminal research initiatives on dogs' biology and psychology.

One of the experiments conducted by the researchers in Bar Harbor showed how early experience is critical to forming relationships through life, even for animals that are genetically ready to love people, as dogs are. In this study, the scientists strictly controlled how much access to human beings different groups of dog pups received. They raised eight litters of puppies in large outdoor fields surrounded by eight-foot-high fences. Food and water were given to the dogs through a hole in the fence, so that they grew up with effectively no human contact. Each week, however, a different pup or two from each litter was taken indoors to be socialized for just one and a half hours per day, and then returned to their mother and littermates. Once the pups were fourteen weeks old, they were all taken indoors and tested for their reaction to human beings.

There were two very striking outcomes to this study. The first is that most of the pups, even though they received only ninety minutes of human contact per day for just one week of life, were as young adult dogs happy around people. This was especially true of the group of pups that received their brief exposure to humans during their seventh week of life. This finding underlines just how easy it is to tame dogs; even a regimen of human exposure that is clearly less than almost all dogs would naturally get in the course of growing up anywhere near people is enough to ensure the pups develop a social connection to people.

As momentous as this first finding was, the second finding is possibly even more important: it is quite possible to raise a dog that will not live comfortably among humans. The group of young dogs who did not meet people until they were fourteen weeks old were — the scientists reported — "like little wild animals." Even given intensive human contact and training later on, for over a month, they showed only slight improvement. They were not tame, and they could not be tamed.

Think about this result for a moment. Tameness — to say nothing of affection toward people — is not innate in dogs, even though their genes make it possible. Rather, this quality is acquired in puppyhood, through exposure. Simply being given the opportunity to see, hear, and smell people at an early age primes dogs to accept human beings throughout their lives. If the exposure occurs at the most sensitive time, then just an hour and a half a day for seven days can be enough. But if that doesn't occur, then dogs' inherited potential to love humans is lost forever. And — although no formal experiment has been done — there is every reason to believe that the same is true of dogs' potential to love goats, or sheep, or even little penguins. Dog pups must get enough exposure to another species in early life if they're to develop a bond with it.

I would dearly like to see how early life experiences shape the loving behavior of dogs, yet I would never dream of conducting an experiment like this one myself. Although the researchers do not state this explicitly, it certainly appears that the dogs they raised without human contact — the ones that could not form relationships with people as adults — were euthanized after the study ended. I think that is unconscionable.

The "experiment" on Middle Island has a happier ending, and one of these days, I really do hope to get to see the penguin-guarding Maremmas for myself, though I haven't had the chance to visit them yet. Fortunately, however, there is no need for such extensive travel in order to see dogs that have formed strong attachments to a species other than our own.

David and Kathryn Heininger are dairy goat farmers in northeast Arizona, close to the New Mexico border. I stumbled across them one afternoon, as I was googling around to see if there might be any livestock-guarding dogs within a day's drive of my desert home. I reached out by email, and the Heiningers responded rapidly with an invitation to visit and see their wonderful dogs for myself. My collaborator, Lisa Gunter, as well as my wife, Ros, came with me, and we drove past Arizona's largest marijuana-growing operation and a variety of shady-looking trailers scattered around the red-dirt hills, before finally reaching the Heiningers' ranch. David and Kathryn describe themselves as hermits, but friendlier and more sociable hermits I could not imagine. And as I soon saw firsthand, their affability extended to their dogs.

The Heiningers' forty goats have round-the-clock protection from three Anatolian sheep-guarding dogs—one of the breeds Ray Coppinger brought back from Europe—and one Old English sheepdog. The Anatolians—Ranger, Mattie, and Kailin—keep the ranch completely free of coyotes, and the Old English sheepdog, Kingman, is supposed to mop up the prairie dogs that slip in because the coyotes are gone. (Prairie dogs are not dogs and do not live on prairies; they are basically ground squirrels, and pose no harm to the goats. But they are a nuisance to have around in large numbers because they eat the already sparse vegetation, and Kingman clearly loved chasing them out of his domain.)

Contrary to what you may have heard, guardian dogs very seldom get into fights with the predators they come up against. Typically, a coyote, wolf, feral dog, or other interloper will simply move on once it realizes that the stock are being protected by a professional livestock guardian dog. This means that these dogs are a form of nonlethal deterrence—something that is very valuable in a world where so many large predator species are endangered.

The Heiningers' dogs are not, as I had been led to expect, aloof toward people. All four of their dogs approach visitors with friendly interest, expecting to be petted. David and Kathryn explained to me that ranchers differ in just how human-oriented they want their dogs to be. Since they live on a fairly small farm where the dogs are seldom far away from people, the Heiningers don't mind their dogs having some sociable interest in humans. But after greeting visitors, the dogs all went back to hang out with the goats. They had all been put with goats when young and had imprinted on them.

David and Kathryn Heininger's goat Tarragon resting with her kid on an Anatolian livestock-guarding dog named Pierre

The dogs did not express their caring for the goats very effusively. The dogs stayed close—typically not more than ten feet away from the nearest goat—but they did not attempt to rub up against or interact with one in any other way. Kathryn described the relationship between the goats and the Anatolians as "like an old married couple," and that seemed like a good way of putting it. There was evidence of concern and caring, but not a lot of open physical affection. Kathryn explained that this was intentional. It is important that the dogs stay close to the goats when the young are born, but dogs that are too excited and want to play with their charges can accidentally harm the kids. Consequently, young dogs are discouraged from having physical contact with the goats.

Though the Anatolians may not have been the most expressive of loving companions, their concern for their goats was very apparent. Most of the time, the dogs just seemed to be sleeping on the red dirt, but if an intruder appeared, they were promptly huffing and puffing and making their presence known. Even crows settling on nearby trees got shown the door pretty quickly. When I tried to return to the humans and goats, having left the group briefly, Ranger, who had been apparently fast asleep nearby, made it pretty clear that I was not welcome. Dave had to intervene and explain to the dog that I was OK and should be allowed back into the fold.

Incidentally, Kathryn mentioned that she and Dave can readily tell from the tone of the bark what kind of intruder the dogs are concerned about—be it crow, snake, coyote, human, or something else. This tells Kathryn and Dave whether the dogs need help and, if they do, what kinds of implements they might need to bring along to deal with the trespasser. Kathryn's comment mirrored almost exactly the claim that I heard from the Mayangnan hunters in Nicaragua, who said that they could tell what kind of prey their dogs were after simply by the sound of their barking. I was fascinated by this new example of the strength and utility of the dog-human bond—and glad not to have occasion to test it.

The example of livestock-guarding dogs shows just how a dog's early life experiences prepare it to love people—or goats, or sheep, or whatever else it grows up with. The internet is awash in tremendously cute examples of dogs that have made friends with ducklings, guinea pigs, rabbits, piglets, a tortoise, a cow, and so many, many others. Dogs reared with cats may even make friends with their traditional feline foes.

The initial programming for affection that dogs receive from their genes appears to be shockingly open. Whereas wolves and other wild animals are suspicious of unfamiliar individuals even from species (including their own) that they have imprinted on early in development, dogs are much more ready to make new friends throughout life. This can be measured in studies that assess the strength of the affectionate bond between two individuals.

I mentioned in Chapter 2 that psychologists who study the development of our own species have found ways of assessing the strength of the connection between child and caregiver — and that one of the most widely used of these tests, the Ainsworth Strange Situation procedure, has been utilized to provide evidence of how our dogs are attached to us emotionally. It also sheds light on how dogs form such bonds in the first place.

As you may recall, in this test a mother (or other caregiver) and her child are introduced to an unfamiliar space. There then follows a series of phases, stretching over twenty minutes in total, during which the child spends time alone with a stranger and then with the mother again. The procedure is intended to mimic the natural flow of familiar and unfamiliar people in a child's life, exposing her to a little mild stress along the way. Children with a secure bond with their caregiver usually explore happily while the caregiver is present, which may include interacting with the stranger. But children become visibly upset, and withdraw into themselves, when the mother leaves. When the mother returns, the securely attached child's happiness is evident, and she returns quickly to contented investigation of her world. Less securely attached children may ignore their parent, fail to explore no matter who is present, show distress even before the mother leaves, or just appear generally stressed throughout the procedure.

When dogs are put through this assessment, they display extremely strong connections to people—especially their owner. In the presence of their special person, they appear confident, but when she disappears, they look anxious—a pattern that psychologists term "secure attachment" when they see it in our own species. This demonstrates just how powerfully connected dogs become to people with whom they share a special relationship. This is a remarkable finding—but even more remarkable is what transpires when dogs who do not have a special human are put through the test.

In one study from the famous Family Dog Project in Budapest, Márta Gácsi and her collaborators tried the Strange Situation test at an anonymous dog shelter. None of the dogs in the shelter had a "primary caregiver," or "pup parent," or "master/mistress," or whatever term you prefer. There were no humans in the shelter with whom they had been given the opportunity to form a meaningful bond, or even socialize with at all, really. The dogs lived in very large groups—up to one hundred dogs—in yards over a quarter of an acre in size. They were fed, and their yard was cleared once a day by a caretaker, but aside from that they had basically no human contact.

The experimenters took thirty of these dogs and gave them ten minutes per day of individual play and handling. One of two women talked to each dog, petted it, and gave it simple exercises and some opportunities for play. Just ten minutes per day, for three consecutive days. Then they put these thirty dogs and another thirty dogs they had not interacted with through a simple version of the Strange Situation test. The person who now had thirty minutes of familiarity with the dog took the role of the mother, and the other experimenter was the stranger. For the thirty dogs that had not received any time with either experimenter, the roles of mother and stranger were assigned at random.

I was very surprised to learn that, when trained observers—who did not know which dogs had received the limited amount of handling and which dogs were completely unfamiliar with both of the experimenters—went through video records of the study and analyzed them the same way that children would be assessed in this task, the observers found clear evidence that the dogs with some exposure to one person were showing attachment toward that person. The handled dogs spent less time by the door trying to get out, and when the familiar person returned after having been out of the testing room, the handled dogs showed more contact-seeking toward her. Overall, the dogs with just thirty minutes of handling were described as using the familiar person as a "secure base," which is one of the signs of secure attachment.

When I read that dogs will start to show an attachment so quickly, I was really quite astonished. It is vanishingly unlikely that the young of our own species could form attachments that rapidly. (Experiments on children raised under conditions similar to those of dogs in shelters are—thankfully—all but inconceivable. Tragically, in some situations—such as the discovery of neglected children in orphanages in Romania at the end of the Soviet era—scientists have been able to document the pitiful and abiding consequences when children are reared without the stabilizing presence of a constant caregiver. The impacts of this deprivation are not so easily remedied for young humans as they appear to be for dogs.)

My past student Erica Feuerbacher (who carried out the studies of dogs' preferences for petting or food, described in Chapter 2) and I rather serendipitously uncovered additional evidence of the rapidity with which dogs can form relationships. We were investigating how the behavior of pet dogs toward their owners was different from their behavior toward strangers. As part of this study, we investigated the reaction of thirteen shelter dogs when they were given a choice between two people. The shelter dogs were included only as a sort of control group—but as it turned out, the results were quite surprising.

These shelter dogs lived in individual pens or paired up with a kennel buddy, and met volunteers and shelter staff every day; also, members of the public interacted with them from time to time. But aside from that, these dogs had no particular special people in their lives.

Erica brought each of these unattached dogs into an unfamiliar room at the animal shelter and gave it a choice between two young women seated on chairs about two and a half feet apart. Both these people were ready and willing to pet the dog if it came close enough; both were complete strangers to the dogs.

Even though the shelter dogs had never met either woman before, within the course of a single ten-minute session, most of the dogs developed a strong preference for one or the other. The preference the shelter dogs formed for one stranger over another was of a similar magnitude to the preference that pet dogs showed for their master or mistress when given a choice, in the same setup, between their owner and a stranger. Pet dogs spent on average a hair less than eight minutes of the ten-minute interval with their familiar caregiver. Shelter dogs were with their preferred stranger more than seven and a half minutes out of the ten.

It is quite astonishing that dogs can form preferences for one human over another so rapidly, though I should warn against the conclusion that these dogs were showing as intense a connection to a stranger after just ten minutes as they would experience toward someone they had been living with for years. When we compared the behavior of pet dogs choosing between their owner and a stranger and these shelter dogs choosing between two strangers, we did notice differences. About half the pet dogs spent every last second of the ten minutes right up close to their owner. None of the shelter dogs spent such a high proportion of the time period with their preferred stranger. And some of the pet dogs actually spent a lot of time with the stranger. A willingness to explore an unfamiliar environment, so long as the caregiver is close by, is actually a sign of a secure attachment. None of the shelter dogs spent much time with the less preferred person.

So there are certainly differences between how a pet dog responds to its owner and how a shelter dog behaves toward the person it prefers out of two random strangers. But the fact that dogs can form preferences for certain humans so quickly is nonetheless very striking and confirms that there is something different about how dogs and humans form relationships. I think it probably has to do with the genes that have been discovered to connect dogs to the exaggerated sociability of Williams syndrome. What these experiments are bringing to the fore is the social fluidity, extroversion, hypersociability—basically the capacity to form loving connections—of dogs. They are so much more emotionally available than our own species or wild animals: and that, I am sure, is a considerable part of their charm.

But whereas dogs form bonds more rapidly than people do, I also suspect they can loosen them more easily.

I don't mean to undermine the notion of dogs' famous loyalty; dogs certainly have put themselves in harm's way and even been killed while defending beloved people. There are thousands of these accounts, some of which are doubtless apocryphal or exaggerated, but some of them are surely legitimate. Skepticism can't stick a pin in all these balloons—there simply are too many of them. For example, Pete, an older rescue dog, died protecting his master and the family's other dogs from a black bear, which they surprised on a hike through the woods one winter morning in Greenwood Lake, New York, in 2018. In 2016, Precious, a pit bull service dog, gave her life to protect her master, Robert Lineburger, by jumping between him and an attacking alligator. Jace Decosse's dog Tank, a pit bull cross, died in 2016 while trying to defend him when assailants attacked him with a crowbar as he slept in his bed in Edmonton, Alberta. All these, and many, many more, are true stories, and just because we are never going to do an experiment to confirm whether a dog is willing to die to save its master does not make them any less convincing to me.

I'm skeptical, however, of tales like those of Lassie in the original 1940 book *Lassie Come-Home*. In the story, this dog trekked hundreds of miles to find her way back to her original human family. If it were really true that dogs would rather die than be rehomed, the millions of people who each year adopt adult dogs that had previously lived with other families could not possibly have the happy outcomes that we know they do.

My own mutt—sweet, simple Xephos—spent the first year of her life with another family before she found her way to ours. The first couple of weeks we had her, she was clearly puzzled and upset. Within a month, however, she was so happy to be with us, no one ever guessed she had not joined our gang as a tiny puppy. She has lived with us for six years now. Recently, just to see how she would react, I said the name we know she had for the first year of her life. "Thyra," I said—rather uncertain what to expect. I received nothing, no response at all. I'm not sure this implies she has entirely forgotten the people she first lived with. I do wonder how she would respond to her first human family if she saw them again. Just because she does not recognize her old name doesn't necessarily mean she wouldn't recognize the people. Charles Darwin pondered this question as he spent five years sailing around the world. He was shocked to find that his dog, Pincher, still recognized him when he came home after such a long time away. According to Emma Townshend, in her beautiful little book *Darwin's Dogs,* it was Darwin's sister Caroline who, in a letter, posed this question: "I wonder if Pincher will be very glad to see you again?" Darwin was so impressed by how Pincher responded that he included the experience in *The Descent of Man:* "I had a dog who was savage and averse to all strangers, and I purposely tried his memory after an absence of five years and two days. I went near the stable where he lived, and shouted to him in my old manner; he showed no joy, but instantly followed me out walking and obeyed me, exactly as if I had parted with him only half-an-hour before. A train of old associations, dormant during five years, had thus been instantaneously awakened in his mind."

I have to wonder what Darwin meant by pointing out that Pincher "showed no joy"—which appears to suggest the dog had forgotten him; but then he immediately adds that Pincher "followed me out walking and obeyed me, exactly as if I had parted with him only half-an-hour before" — which clearly states that the dog had remembered him. I think it is possible that dogs may remember people for many years, and yet perhaps the emotional bond fades with time. It seems to me that the rapidity with which dogs can form new bonds implies that old bonds must fade, but presently this is entirely speculation; no studies that I am aware of support this idea.

In any case, I prefer to focus on the happy beginnings of loving relationships rather than their possible demise.

I still remember vividly how my mother and I went to the animal center of the Royal Society for the Prevention of Cruelty to Animals on the Isle of Wight, to see if there might be a dog there to suit our family. We had promised my younger brother, Jeremy, that we would not bring one home without first giving him a chance to vet (pardon the pun!) our selection. But when we came to the kennel that held Benji, he made such a strong case that he was our dog that we just could not bring ourselves to drive off and leave him there. My mother handed over five pounds, and Benji was ours. Jeremy was annoyed with us, but Benji worked his magic on him too, and Jeremy quickly became convinced that this was truly our dog.

The experience my family had forty years ago is far from exceptional. Many people report that they didn't really choose their dog — their dog chose them. They cannot pin down how the dog did it, but something in her demeanor — the look in her eyes, her body posture — somehow convinced the humans that this dog had been theirs all along, and by taking the dog home they were simply completing a connection that already existed. This ability to persuade people to take them in and protect them must surely be a key component of dogs' success in human society.

The rapidity with which dogs develop an affectionate interest in people is quite startling, but it is part and parcel of how open their programming for forming bonds with members of other species really is. Just because dogs have the coding for this wonderful openness, however, does not mean that they always get to use it.

Although having the right genes is crucial to dogs' loving nature, it is far from the whole story of who they are. Individual dogs grow up to be a range of things: the loving partners that share our homes, dangerous and aggressive beasts that people have no desire to come near, and even wild animals that are afraid of people and never seek their company. Life experience is the clincher. Even clones, built quite literally from the exact same genetic materials, born from the same mother, growing up in the same human household, can nonetheless develop substantially distinct personalities. (How I would love to be able to give doggy personality tests to a hundred sets of clones!)

Livestock-guarding dogs are the most widespread example of dogs that show affection toward species other than our own. This demonstration of how dogs can grow up to love other beings is both widely dispersed around the world and quite ancient. The experiment from seventy years ago at Bar Harbor demonstrates that the right early life experiences are crucial if dogs are to grow up loving people. That experiment was directed by two scientists who were both geneticists, John Paul Scott and John L. Fuller, but late in life, when Scott was invited to write a reminiscence about what that major project had taught him, his succinct summation was this: "genetics does not put behavior in a straitjacket."

The science of genetics has advanced by leaps and bounds since the days of Scott and Fuller, while the techniques of behavioral study — which really amount to not much more than systematized observation — have not changed nearly as greatly. Perhaps that is why, when I speak to audiences made up of the dog-loving public, people are very ready to be told that there are genes that make dogs loving, whereas they seem more surprised by the much-longer-known facts of the role of a dog's environment in determining its character. Old science can still be good science, I tell them. And in fact, it's the old science, the study of how the world around an animal shapes its character, that can have the bigger impact.

It is true that genetics and environment are equal partners in defining individual dogs, as they are in all of biology. But we do not have equal control over them. Genes we are stuck with. Sure, with present technology, at phenomenal expense, tiny changes in an individual's DNA can be made—but genetic engineering is still very much the toolkit of the future, not the present. The environment our dogs live in, on the other hand, is completely under our control.

We create the worlds our dogs are born into; we can change those worlds if we want to. By doing so, moreover, we can enable them to live their best lives with us. Immense responsibility comes with this great power.

7

DOGS DESERVE BETTER

THE BEAUTIFUL BOND we share with our dogs today is only possible because, sometime during the past fourteen thousand years, a small change occurred in their genes. This tiny mutation turned dogs from guarded beasts inclined toward only a few strong connections with other members of their species into adoring creatures who readily and rapidly form emotional bonds with almost any type of animal.

But dogs' genes only make it *possible* for them to love. The world they grow up in is what actually causes them to love. In a sense, we ourselves turn dogs, one by one, into "man's best friend."

For over half a century, scientists have been aware that a process takes place early in each animal's life that determines the kinds of beings it will seek for emotional connection for the remainder of its time on earth. Dogs love us because we are there for them during this sensitive period.

It is only right that we be there for them throughout the rest of their lives too.

As individuals and as a species, our dogs have pledged themselves to us. They gave up their ancestors' fearsome jaws and brilliantly coordinated hunting abilities. They gave up their forebears' tight-knit family life to seek bonds with creatures beyond their own kind. They abandoned a life of roaming and hunting. They exchanged these things for the chance to partner with humans. Ours is a covenant never spoken aloud, but nonetheless sincere and binding on both parties.

In exchange for their deep and enduring affection, dogs trust that we will care for them. That we, the puny, hairless, but very clever apes that we are, will use our smarts to ensure their well-being.

I have never heard anyone say their dog did not keep up its end of the bargain. The loyalty of dogs' affection is the stuff of legend. But we, I'm sorry to say, have not always kept our side of the deal.

Of course, many dogs receive terrific care. But too many others — millions in the United States alone — do not. We don't give them the sort of life they deserve, that they trust us to provide, and that they often mutely plead for. Too many of the ways in which we structure the lives of our dogs are archaic and do not reflect the latest science; many of our most widespread practices are positively barbaric.

Luckily, the science is every bit as clear about how we should be managing the canine-human partnership as it is about the underlying nature of that relationship. From affirming the importance of our steadfast presence in dogs' daily lives, to supporting specific training techniques, and even demonstrating the benefits of certain kinds of physical touch, cutting-edge research is rife with lessons for how we should modify our behavior based on the theory of dogs' love. Science has not only revealed the emotional core of the dog-human relationship; it also contains concrete lessons for how we can ensure our own dogs' emotional well-being, if only we care to look.

And we must find ways to improve. If there is one thing that my journey into the heart and soul of this incredible species has taught me, it is that it is incumbent upon us not simply to understand our dogs' emotional needs but also to do something with this information. To put it bluntly: we can do better for our dogs. The rich science and history behind their loving nature make it clear that they deserve it.

Our failure to care for our dogs adequately is usually due to a faulty understanding of these animals and their needs.

If you believe—as some people sadly still do—that dogs are still essentially wolves, then you will adopt an attitude of wary alertness toward your pet. You will be perpetually braced for your gentle hound to turn into a monster, and you will be easily convinced, when supposed experts tell you to assert yourself with whatever force you can muster, to compel this unpredictable beast to accept your dominion.

If, on the other hand, you accept the argument I have developed here, that dogs are differentiated from their wild cousins by their capacity to form strong emotional ties with members of other species, including, in particular, ourselves, then you will seek a way of living with your canine companions that enables you to coexist lovingly and peacefully with them.

Of course, just because two individuals are capable of having a loving relationship does not guarantee that problems will not crop up from time to time. And when these life partners belong to different species, even the best of intentions cannot protect them from occasional misunderstandings and, shall we say, creative differences.

Problems in a canine-human relationship are inevitable. But the way you understand these problems is up to you—as is the way you respond to them.

The view that dogs are basically unreformed wolves encourages people to take a terrible approach to such problem solving. This outlook allows dog owners to justify exerting physical strength to correct a perceived imbalance of power between them and their dogs, while neglecting love's role as a powerful motivator for these animals. This is a tragic misreading of dogs' true nature, and it can have real consequences. If you adopt this way of thinking, the odds are good that you will eventually cause physical or psychological harm to your dog.

This approach has been around for a long time, sadly, and its influence has proved both insidious and incredibly tenacious. One of the most popular books on living with dogs, *How to Be Your Dog's Best Friend,* by the Monks of New Skete, was originally published in 1978. It argued that dogs, because they are descended from wolves, can understand life only as it is lived in packs, and furthermore, that a pack is a hierarchically organized community in which competition for status forever flickers below the surface, occasionally bursting into outright conflagration. Many of the problems that people have with dogs, the monks argued in 1978, stem from a failure to recognize these essentials of dogs' social life, and such problems can be remedied by asserting one's "dominance" over the canine members of the household.

Perhaps the most infamous of the monks' suggestions was the disciplinary measure that they termed the "alpha roll." In this maneuver, the dog owner was encouraged to emulate the discipline meted out by an alpha wolf toward a subordinate animal. Specifically, the monks instructed owners to roll the dog abruptly onto its back and grasp it firmly by the throat, while scolding the dog forcefully.

Now, before we subject the monks to equally harsh discipline, I should mention that there is much that I admire in their overall approach to dog-human relations. The Monks of New Skete reach beyond just the mechanics of training a dog to emphasize the importance of paying attention to its comfort and contentment. They highlight the dog's social nature and encourage owners to include their dog in as many aspects of life as is practical, advice that I heartily endorse (short of getting your dog a fake service-animal vest). Their focus on heightening humans' empathy for dogs is laudable, and I wish more people followed their recommendations in this regard.

It also is only fair to note that the general approach the monks outlined in 1978 was not inconsistent with the available science of the time. Research on the social lives of wolves was in its infancy back then, and study of the psychological differences between dogs and wolves had not even started. Early reports of how wolves behaved around one another were almost entirely based on groups of largely unrelated animals brought into captivity together. These studies did find very high levels of aggressive status competition among the wolves the researchers examined.

But while this research provided an accurate depiction of wolves' behavior in captivity, as a reflection of inter-wolf relations in general, it was woefully incomplete. Thanks to the subsequent work of David Mech of the United States Geological Survey and the University of Minnesota, and other field biologists, we now know that wolf packs in the wild are simply nuclear families. The so-called alpha male and female are usually the parents of the other pack members. Relationships within the pack are certainly hierarchical, but no more or less so than in a human family. The alpha animals show more loving concern than violence toward the other pack members, and free-living wolf packs are characterized by very low levels of aggression. If wolves in the wild experience chronic tension with other group members, usually some of them pack up their bags and leave home. Since captive wolves cannot do this, they tend to show higher levels of aggression toward one another. If you were trapped in a prison cell with an estranged brother or sister (or a complete stranger, for that matter), you might respond similarly.

What's more, we now know that wolves and dogs, closely related as they undoubtedly are, operate within very different social structures. As I'll explain in more detail later in this chapter, dogs and wolves have very different dominance hierarchies, which cause them to relate to members of their in-group in disparate ways. These different ways of relating to members of their own species (and in the case of dogs, other species as well) make it very difficult to extrapolate useful prescriptions for dealing with dogs based on observations of wolves' social behavior.

None of that was known in the 1970s when the monks were working on their book.

Commendably, in a revised edition of *How to Be Your Dog's Best Friend,* published in 2002, the monks disavow the "alpha roll." "*We no longer recommend this technique,*" they state emphatically, "and strongly discourage its use to our clients." It is rare that people have the confidence to go into print flatly contradicting something they previously promoted, so the monks earn considerable kudos from me for undertaking such a radical revision—although, since the individual authors of the book are anonymous, it's possible that the monk who wrote the disavowal of the alpha roll in 2002 was not the same brother who had first recommended it thirty-three years earlier.

Alas, not everyone has caught up with the monks. Other high-profile dog trainers still unapologetically promulgate techniques that, though they were arguably state-of-the-art in the late 1970s, today are widely recognized as cruel and indefensible. Many of today's best trainers advocate methods that do not depend on force but instead concentrate on positive consequences and gentle leadership. Trainers and teachers such as Victoria Stillwell, Karen Pryor, Marty Becker, Ken McCort, Jean Donaldson, Chirag Patel, Ken Ramirez, and many, many others are up-to-date on science and know that coercion, pain, and punishment are not the right foundation for building a relationship with a dog.

Unfortunately, in the realm of television—still the dominant medium in our culture—expertise is not always the most valued currency. Charisma and screen presence are far more important; everything else can be fixed in postproduction. As a result, some wildly distorted and unethical advice is given to dog owners.

Some baseless guidelines that TV dog trainers advocate are just silly—such as making sure that you eat before your dog does, or that you walk through a doorway in front of it. But others are not so harmless. We see animals strong-armed on "slip leads" (basically a noose), kicked, "flooded" (exposed to extremely stressful levels of a stimulus they cannot escape), and treated in many other inhumane ways.

Of course, these brutal methods allow trainers and dog owners to bring the misbehaving animal quickly to submission—but at what cost? The fallout from these coercive approaches is not presented on TV. Either the evidence is left on the cutting-room floor or becomes apparent only after the TV crew has left town. A dog's behavioral problems will grow worse than ever as it becomes chronically anxious and afraid of people.

The trainers who advocate for these inhumane measures defend them on the basis that it is essential for a human to be the "pack leader" of his dogs. People are told that their canine companion is a beast programmed to be forever in competition for "top dog" status, and humans have to do whatever is necessary to disabuse a dog of the notion that it can outrank a human in any way. This has led to an enormous amount of confusion surrounding the concept of dominance, and no small amount of collateral damage—so it is worth pausing here to consider what dominance does and does not entail when we are talking about dogs.

In animal behavior, dominance simply refers to a social situation in which certain individuals routinely gain preferential access to constrained resources. This could mean that, when a limited amount of food is available, particular animals get to eat first. Or when a female is in heat, certain males get to mate with her first (or are the only ones to mate with her at all). Or when the weather is inclement, particular individuals get preferential access to shelter.

I had taught basic principles of animal behavior to legions of students before I thought about dogs from a scientific point of view. As a result, I knew a thing or two about dominance (as it relates to animals, I mean). But when I started to talk to groups of people interested in dogs and dog training, and heard them talk about this concept, I felt very confused. Inspired by what they had seen on television, many people I met talked of dominance in a way that suggested world domination or even a dominatrix out of a BDSM scenario. This did not jell at all with my scientific understanding of dominance.

For one thing, not all animals experience dominance in their relationships. If we confine our discussion to Carnivora (carnivores)

—the order of animals to which wolves and dogs, along with a great many other predators, belong—it keeps things a bit more straightforward. Some members of the order Carnivora show dominance, but others do not. Leopards and tigers, for example, are not very social animals, and so dominance means little to them. Even some social animals, such as lions, do not show dominance, though they can be ferocious. Clearly, ferocity and dominance are quite different concepts.

In any event, most species of carnivore are social, and most show some level of dominance in their social structure. But the style and extremity of the pecking order vary by species. For instance, hyenas are an example of a species that exhibits what biologists call "linear dominance." The top-ranking hyena has priority over the second-place hyena; the second-place hyena has dominance over hyena number three, and so on down the list, to the last hyena—a sorry creature that has the least claim on any resources in short supply.

Some other species of carnivore—wolves among them—show what behavioral biologists have labeled "despotic dominance." In this form of social organization, one individual (or pair of individuals) makes all the decisions and the rest of the group just goes along for the ride. In wolves, for example, the alpha male and female—who, remember, are just the parents—make the decisions. The other group members, their pups, follow along.

If you are wondering what these different forms of dominance feel like, you probably already know. Human organizations provide opportunities to see different types of dominance in action, whether despotic dominance (the boss tells everyone in the office what to do) or linear dominance (the boss gives orders to the second in command, and so on down the line). Some human communities don't have clear patterns of dominance: for instance, groups of friends, clubs committed to hobbies, and so on. We humans are certainly flexible social creatures.

In this way we can see that a wolf pack is not so different from a human family. Parents in most human societies that I know of are dominant over their children because these adults are responsible for most of the big decisions, such as what to eat, when to eat, where to live, and so on. This does not imply that parents are forever beating their children into submission. At least it shouldn't.

The crucial thing to realize here is that — as David Mech, the scientist who pulled back the curtain on wild wolves' social lives, observed — dominance does not have to imply coercion.

Dogs' social structures aren't nearly as flexible as ours. In fact, dogs' rigidity in this regard may surprise you. It certainly shocked me when I first read the research about dogs and dominance. Not only are dogs more prone to hierarchical social organizations, with clear dominance relationships, than we are, but they are also more hierarchical than their supposedly dominance-obsessed ancestors, the wolves.

Researchers at the Wolf Science Center in Austria have carried out studies into the degree of dominance dogs show among themselves. As you may recall, the staff at this research facility raises groups of wolves and groups of dogs outdoors, under conditions as close to identical as they can manage. In standardized tests of how groups of individuals behave when confronted with a resource — such as a piece of food — that they cannot share, the researchers found that dogs are actually significantly more hierarchically organized than wolves.

In one beautifully simple experiment, Friederike Range and her colleagues at the Wolf Science Center offered pairs of wolves or dogs a pile of food. Range's team offered each pair of animals a different-sized food pile. This pile had to contain just the right amount of food — large enough to be shared, if that was what the two individuals wanted to do, but small enough for a dominant animal to monopolize, if that was its desire. Range and her collaborators also did the exact same thing with a large bone — again, it had to be big enough for two animals to chew it together, but small enough for a single animal to run off with it and defend it, if that was its will. Range and her team then observed what happened: did one animal keep the other away, or did the two wolves or two dogs share nicely?

The wolves in this experiment were generally inclined to share. Of all the pairs of wolves that the research group tested, in less than one in ten did one wolf claim the food and prevent the other animal from eating any.

For the dogs in the study, the picture was quite different. In about three-quarters of the tests, the dominant dog prevented its partner from eating anything. The issue was not that the dogs were more aggressive than the wolves: dogs and wolves growled and grumbled at each other at similar rates. But when a dominant dog complained at a subordinate dog, the subordinate animal quickly withdrew and gave up trying to feed. In wolves, whatever grumbling there was went back and forth between the two animals, but it did not prompt either to stop eating. Dogs, it seems, are considerably more sensitive to other dogs' claims of dominance than wolves are.

Many other studies show that dogs have steeper social hierarchies and are more prone to (and sensitive to) displays of dominance and monopolization of important resources than are wolves. If this seems paradoxical, it is only because we frequently confuse dominance with ferocity. Wolves are ferocious: they are big, powerful, fierce predators. To be on the receiving end of a wolf attack would be a truly terrifying experience — quite possibly the last experience of one's life. Dogs, of course, are less ferocious: they are smaller, punier, and generally less fierce. That's not to say that I'd welcome an attack by a dog; but put simply, a species' level of ferociousness has no bearing on its level of dominance.

Wolves and dogs differ in their sensitivity to dominance because they lead such different kinds of lives. When we think about how wolves and dogs live, the canine disposition to dominance makes perfect sense. Wolves survive by hunting live prey — prey that is often larger than they are, and that is definitely trying to avoid becoming a wolf pack's dinner. An individual wolf cannot usually bring down a bison or a deer or any other large prey animal that wolves feed on. The wolf depends on cooperation from the rest of its pack to achieve the kill. And once the deed is done and the dead beast is on the ground, there is a lot more meat available than one wolf could possibly eat. The rest of the pack is the dominant wolf's family, and it depends on them in order to hunt successfully. Consequently, that wolf has nothing to lose and everything to gain by sharing the spoils of the kill.

Wolves, in short, live in cooperative packs whose members depend on one another for survival. Numerous studies show that even though their social structures are hierarchical, they are nonetheless capable of high levels of cooperation.

Free-living dogs live quite differently. I saw this firsthand a few years ago, when I visited Nassau in the Bahamas. This has to be one of the finest places on earth to quietly observe free-ranging street dogs, and probably one of the best places to be one. The weather is extremely conducive to outdoor living. The local culture is also unusually accommodating to homeless dogs, as I quickly discovered upon my arrival.

During the course of an afternoon spent exploring the back streets of Nassau, under the guidance of an enforcement officer from the Bahamas Humane Society, I saw some incredible displays of peaceful coexistence between free-ranging dogs and their human neighbors. Dogs cheerfully wandered the roads, minding their own business, as drivers carefully navigated around them, going extra slow to avoid hitting them.

Humans help the dogs of Nassau in other ways as well. Standards of garbage collection in the Over-the-Hill district—the side of Nassau that tourists don't usually see—leave something to be desired, and there were piles of trash at many street corners. I watched as a rather mangy, orangy-brown small dog worked his way through a garbage dump that had accumulated at a twist in the road. He stuck his snout as deep as he could into a discarded box of KFC. This dog certainly didn't need any help getting the food out, so he was not motivated to share what he had found and justly claimed as his own. He even growled at me to make sure I understood this box was his.

Because of the unique conditions in the Bahamas, dogs there lead lives very different from those of their cousins in the United States. But in one respect, at least, they are the same: as animals that live primarily as scavengers, free-ranging dogs have little reason to cooperate with one another in order to find and eat food, and they have every reason to monopolize what they have found. In a way, the same evolutionary pressures that caused this orange-hued Bahamian street dog to bare his teeth at me have caused dogs as a whole to be much more hierarchically sensitive animals than wolves are.

Street dog in Nassau

The acute levels of dominance that dogs display in their relationships with one another, and their exquisite sensitivity to hierarchy in social situations, have profound implications for their lives with us. For one thing, regardless of whether you brutalize your dog using choke chains and alpha rolls or are as gentle as you would be with one of your own children, your dog knows you are the boss. You are the one who makes food magically appear—food that is locked away in cans in cupboards, bags in the fridge, and numerous other containers that are simple for us to open, with our opposable thumbs, but which remain mysteriously closed off to most dogs. You are the one who decides when your dog will leave the house, and which way the two of you will go. You even decide when and where it is appropriate for your dog to relieve itself, whom your dog may have sex with, and even whether it has a sex life at all.

For all these reasons and more, it is clear to your dog that you are the leader in your relationship.

You may not be aware that, by controlling what and when your dog eats, you are exerting dominance over your dog—but all the available research indicates that your dog is acutely aware of it. Dogs understand that whoever controls the resources must be the boss. You may, as I advocate, use food treats, clickers, and gentle encouragement to get your dog to walk nicely alongside you, rather than (heaven forbid) a prong collar or other torturous instrument. But whether you use positive or punitive methods, these techniques all achieve the same aim—getting your dog to walk with you the way you want her to—and if you achieve that, then you are asserting dominance over your animal. Frankly, if your dog is going to live in human society, you need to be dominant over her. Dogs are not psychologically equipped to make the decisions in your family.

You, and only you, can decide what form that dominance takes. To be the senior partner in the relationship, you don't have to shock your dog, jerk her around on a choke chain, or kick her soft underbelly. Your dominance resides in your control of resources, and you can express it by showing compassionate leadership rather than barbarism. As any parent knows, love and dominance are not incompatible. Dogs understand both. They deserve leadership based on compassion, not aggression.

Just as dogs understand and even expect dominance from humans, they also crave social contact. It is literally in their genes to need relationships with other beings. They need to play friendly games; they need to be close to loving people.

Just how close dogs want to be to the people they love depends greatly on the individual. For instance, my Xephos craves touch, but only touch: touching my feet when I'm at my desk or in bed, or just next to me on the sofa. She really hates being lifted up and cuddled, and is ambivalent about a full-on hug when she is on the ground—that seems to depend on her mood at that moment. Some dogs love being picked up off the ground and fully encased in a beloved person's arms; others don't seek continuous contact and are happiest with proximity alone.

There has been some debate about exactly how dogs like to be touched. The Canadian writer Stanley Coren has suggested that dogs do not enjoy being hugged. In a blog post he described an analysis he carried out of photographs people had posted online of their dogs being embraced by a human. According to Coren, the dogs looked stressed in fully 204 of the 250 photographs he uncovered. He advised his readers to "save your hugs for your two-footed family members and lovers."

I feel Coren is overdoing it a bit here, although he makes a good point—people should pay attention to dogs' responses to physical contact, and not simply assume that what makes us feel good feels good to them too. When thinking about how much physical contact is enough (or too much), the crucial issue is to attend to your dog's response. The old adage "Different strokes for different folks" offers a good perspective for understanding dogs as well as people.

One thing is certain: though every dog is an individual, with a personality of his own, which we must learn to understand and respect, all dogs crave warm, loving relationships. We owe it to them to satisfy that modest requirement.

Too often, we fail dogs in this respect. The cruelest thing you can do to a highly social being is shut him up all day where he cannot interact with anyone. Yet that has become the norm for dogs in first-world countries. We love our dogs for their warmhearted nature, yet we say goodbye to them at seven-thirty in the morning and, if they're lucky, we get back to them ten or eleven hours later. Sometimes people nip home after work to let their dog have a quick bathroom break before locking him up again, so they can go off to socialize for the evening with their human friends. What kind of life is that for a dog? Ten hours alone, ten minutes of social interaction, and then another four or five hours alone before their human comes home, collapses, and quickly falls asleep.

In Sweden, the law demands that dogs receive regular social interaction at least every four to five hours. I think this is an excellent principle. If you cannot get home to your dog during the day, then either you should find other ways for your dog to have social contact or you shouldn't have a dog.

Dogs gain social benefits from other household members besides their owners, of course. A well-raised puppy will welcome the companionship of her own species, and even will get some satisfaction from being around cats and other animals—particularly if the dog in question has been introduced to these beings during that critical period early in life, when it learns which kinds of beasts to make friends with. Truly, a whole range of creatures can serve as social partners for dogs and alleviate canine loneliness.

And of course, there are a range of solutions to dogs' epidemic of loneliness other than bringing home another pet. Staying home with your dog for at least part of the day may be an option. That's what I do, but I know I am very privileged to have so much flexibility in my professional life. On the other hand, taking your dog to work is becoming an option for more and more people today; dog-friendly offices are a welcome trend in the United States. Since most dogs are so quick to make friends, you also could hire someone—or prevail upon a friend with a less crazy schedule—to pop around and chat with your dog every day. Maybe they could get a coffee or lunch together. A well-run doggy daycare is another excellent option, and one that many responsible dog owners utilize.

One way or another, however, dogs' open, loving personalities require attention as much as any physical need. Most people wouldn't think they could get away with not feeding their dogs, or not letting them relieve themselves. But keeping dogs shut up alone for long periods of time may be the cruelest routinely accepted thing we do to them. And it has real consequences – for us and for them.

Vast numbers of dogs cannot cope with their crushing solitude, and act out in all sorts of ways, from barking to chewing up furniture to inappropriate soiling in the house, among many other symptoms of loneliness. We label these signs of distress "separation anxiety" and treat them with drugs or behavioral interventions. They have become the most common behavioral problems reported to veterinarians and animal behavior specialists, affecting about one dog in every five.

While I was in Nassau, I visited with William Fielding, a social scientist at the College of the Bahamas. Fielding has given the exact same questionnaire about the dogs that roam the streets of Nassau to native Bahamians and to tourists who visit that beautiful archipelago on cruise ships. Most of these visitors are from the United States. The questionnaire asks what a person thinks is the kindest, and the cruelest, thing you could do with your dog while you are out at work during the day. The Americans answered that a dog must be safely secured in a home while the humans are away. Local Bahamians, on the other hand, were much more likely to respond that a dog should be allowed out of the house if there is nobody home to keep it company.

There is no single right answer to this question, in my opinion. The American respondents are of course correct – to let your dog wander the streets unsupervised is to invite disaster. A dog could be run over, attacked (I saw three schoolchildren make to kick a dog on the street before we yelled at them to leave the dog alone), catch a disease from another dog, or fall victim to any number of other sad events.

But the Bahamians have a very good point too. Because dogs are social beings, it is undoubtedly cruel to lock them up alone in a house for the whole day. Dogs deserve better opportunities to fulfill their social destiny, and we are more than capable of satisfying that need.

I feel for the many dogs that live in homes where they are not given the loving contact that they crave — but the plight of dogs in shelters makes me so sad, it is hard for me to write about it.

The shelter is the dirty underbelly of our lives with dogs. We say we love dogs, yet every night in the United States around five million of them sleep on concrete floors behind iron bars. Although the situation has been improving over recent decades, shelters still admit somewhere over four million dogs a year. Nearly three-quarters of those dogs either get adopted or returned to their owners. But that still leaves about a million dogs, which are either euthanized or become long-term stays in the shelter system. Neither is an acceptable solution to the problem of homeless dogs.

Most shelters in the United States were built when their purpose was solely to provide brief respite to dogs that had strayed from home. Rex might have spent a few days or at most a couple of weeks at the shelter as he waited to be collected by his owners, or adopted into a new family. If neither happened, he was usually euthanized. One way or another, no dog stayed in a shelter for very long.

In many ways, the shelter system in the United States has greatly improved. Today many more dogs make it out of shelters alive than they used to. But as shelters have found ways to reduce the amount of killing that occurs within their walls, lengths of stay have metastasized.

More and more shelters have adopted the position that they will not euthanize a healthy dog, thanks to a movement begun about two decades ago to end all euthanasia of healthy dogs in shelters. The intentions behind the no-kill movement are undoubtedly noble, but good intentions are no protection against the law of unintended consequences. While I honor and respect the movement's commitment to canine well-being, I worry about the unintended consequences of keeping dogs for extended periods — in some cases, the rest of their natural lives — in conditions that were never designed to be anything more than temporary.

I share the desire not to kill healthy dogs, but I also know that simply locking up millions of dogs and throwing away the key is not an acceptable alternative to euthanasia. When a shelter decides to no longer euthanize any except terminally sick animals, its cages gradually fill up with dogs that cannot find homes. There can be many reasons why people don't want to adopt certain dogs. Even if we think some of these may be superficial and regrettable — such as fashions for dogs of certain colors and shapes — that doesn't change the fact that we cannot force people to adopt dogs when they do not wish to. Since no dog's behavior improves while it lives in a kennel, over time the dogs in no-kill shelters become less and less attractive to prospective adopters. These shelters become, in effect, canine warehouses.

Dogs in long-term kennels at a no-kill shelter

Some nations have adopted laws that prevent shelters from euthanizing healthy dogs, but this sort of legislation sadly is not enough. I tried to visit a public shelter in Italy, one of the nations with a policy of this kind, but was barred from entering. The fact that the shelter would not even allow a professional visitor to see how they keep their dogs is enough of a signal that the situation there must have been pretty desperate.

I was able to visit a private shelter near this public one, and I have to say the conditions were among the saddest I have seen anywhere. I am not going to name names here because I know that the people running this shelter were sincerely trying their very best to provide a loving environment under the most difficult circumstances. But having looked these dogs in the eye, to me, the long-term holding of dogs where the resources to care for them properly are not available is just as sad as painless euthanasia.

The news from Italy is not all bad, however. Recent research there has illuminated one way in which we can make long-term shelter stays more bearable for dogs. And given what we know by now about dogs' loving nature, it should come as no surprise that the solution has to do with simple human presence.

A large team of scientists from several Italian universities, led by Simona Cafazza, investigated the welfare of nearly one hundred dogs living out their lives in shelters in the Lazio region of Italy. They found that the only intervention that improved the welfare of the dogs was daily walks with people. Although the study did not prove whether it was the exercise component of the walks or the human company that led to the improvements, Cafazza and her colleagues did compare the dogs that got daily walks with people to dogs that were allowed into larger runs to exercise on their own. Only the dogs that were walked with people showed the benefits in welfare.

Overall, Cafazza and her colleagues were skeptical of the value of their nation's no-kill legislation. It had failed to control the free-ranging dog population, yet led to large numbers of dogs spending their lives in shelters where their needs could not be properly met. The researchers noted that, in the region they studied, eleven thousand dogs lived in shelters; the majority stayed in kennels for their entire lives. Cafazza and her colleagues concluded, "Given that, in Italy, we have decided for dogs that life-imprisonment is better than painless euthanasia, it is our ethical duty to guarantee them an adequate level of welfare. It is evident from the scientific literature that this is not the case."

Different nations, and different regions within nations, face different challenges with their shelter-dog populations. In the United States, we have every kind of shelter problem—and also many of what must be the most wonderful animal facilities anywhere in the world. I have visited shelters with lovely, bright rooms, with cheery paint on the walls, natural light, gentle background music, and charming staff; only the food catering puts me off from settling into such places myself. But I have seen nightmarish shelters too, and the pitiful sight of so many sad, sickly dogs was intensified by interminable barking and the stench of doggy diarrhea.

The fate of a dog in a shelter in the United States depends on many factors. Shelters in the Northeast euthanize fewer dogs because they actually lack numerous canine customers; the widespread sterilization of pet animals in that quadrant of the country has led to a significant reduction in the number of dogs ending up without human homes. Shelters in the Southeast, on the other hand, still contain large numbers of surplus dogs, as do many shelters in the West.

The best shelters provide a peaceful way station for dogs awaiting new homes. These facilities have professional staff on hand to teach their guests behaviors that will help them get adopted, as well as useful life skills for cohabiting with humans. These tend to be smaller, bijou shelters, often supported by wealthy private benefactors. At the other end of the spectrum, there still are shelters that euthanize most of the animals that have served out their fourteen-day holding period. These operations tend to be run by the government of whichever municipality the poor animals happened to be picked up in. Sometimes these diverse types of shelters can be situated just across the street from each other.

I'm not casting any aspersions here. I fully recognize that local governments have many demands on their limited resources. I understand that animal care and control cannot rate higher than funding schools or senior centers or many other civic obligations.

I do believe, however, that dogs deserve better from most of the shelters that take them in. Even the most underfunded facilities can do a better job of reciprocating dogs' love for humans and helping them to better express their love for us in return. In so doing, shelters can help the dogs get adopted quickly, without spending any more time than they need to between homes. This is good for the dogs, and it's good for the shelters too. My students and I have been seeking to help shelters realize this ambition.

When she was working on her PhD with me, Sasha Protopopova —now a professor at Texas Tech University—started a line of research that she continues as of this writing. The goal that Sasha set for herself was to find ways to improve the behavior of shelter dogs in order to make them more adoptable. Her aim was to do this without imposing any additional burden on the staff of the shelter—or, if that was impossible, then at least to help these dogs without requiring any additional personnel with expertise in animal training.

To start with, Sasha spent an interminable summer directing undergraduate helpers in a field research project at a shelter in North Florida, run by the local government. The undergrads would stand in front of each dog's cage for sixty seconds, using a video camera to record whatever the dog chose to do. The time limit was intentional: most people don't spend more than a minute looking at a dog before deciding whether to find out more about it or move on to the dog in the next kennel. Sasha ended up with thousands of these brief videos, capturing the behavior of hundreds of dogs in this common adoption scenario.

The next step was to go through every last second of every one of these videos, noting precisely what each dog did. Did it wag its tail? Bark? Poop? The list of possibilities ran to over one hundred discrete behaviors.

By the end, Sasha had a record of a large number of dogs reacting to a stranger who looked at them briefly. So far as a dog could tell, any such person might have been a prospective adopter. It was as if we had a massive compendium of these dogs' elevator pitches for new human homes.

Sasha took this extensive and intensive record of the dogs' behaviors and compared it to the shelter's records for those dogs. Some got adopted quickly—others languished a long time. By comparing the behavioral analyses to the length-of-stay records, she was able to identify the things dogs did that got them out of the kennels quickly and the things that tended to leave them waiting forlornly for a new human home.

Her first finding was not unexpected because it appeals to common sense and had been observed repeatedly in other studies: if you are cute, it doesn't much matter how you behave. Physically attractive dogs, like puppies and toy breeds, could do whatever they wanted and still get into human homes quickly.

But for the rest of us—sorry, I mean, the rest of them—behavior was indeed crucial to their fate. It turned out that one major turnoff for would-be adopters was leaning. Leaning or rubbing against any part of the kennel really worsened a dog's chance of adoption. Too much movement was bad too; people apparently don't want to adopt dogs that fidget, pace back and forth in their kennel, or jump up and down. The dogs with the best chances for adoption came to the front of the kennel and looked interested in their visitor, in an alert but not overwhelmingly energetic manner.

In an ideal world, shelters would be able to engage professional dog trainers to erase the undesired behaviors, so that any dogs exhibiting them could have their manners corrected and quickly find their way into new homes. Sasha recognized, however, that the majority of shelters, at least in the United States, simply do not have the resources to train their staff to become behavioral experts, nor to hire such people.

As a workaround, Sasha and I thought about techniques that might nudge the dogs' behavior in the right direction without requiring any special expertise. We settled on a path that had been blazed decades earlier by that great Russian physiologist and founder of animal psychology, Ivan Petrovich Pavlov. Sasha was born in Russia, so I'd like to ascribe at least part of the inspiration for this solution to her early life in that country. She left Russia at the age of eight, however — so unless they teach a lot more animal psychology in elementary schools over there than we do in the West, this surmise may not be entirely accurate.

Anyway, we were inspired in part by Pavlov's demonstration that animals can detect a signal that something important is about to happen. The bell (or rather, buzzer) in his now-legendary experiment warns the dog that food is coming; it responds by salivating. This type of conditioning has one advantage over other, more modern forms of animal training: you don't have to pay any attention to the animal. Obviously Pavlov and his students were interested in what their dogs did, but to carry out their procedure, these researchers did not actually need to watch the dogs.

You can easily condition your dog to expect food when you ring a bell without having to look at him. Just ring the bell, deliver the food, and the animal will take care of its own behavior. Of course, if you want to know *how* his behavior has changed, you will need to open your eyes and look, but unlike standard reward-based animal training, where a trainer observes carefully and then gives a treat for the appropriate behavior as soon as it takes place, Pavlovian procedures are far easier to administer. Ring the bell: give the food. Just do that at regular intervals, and the magic will happen.

The price you pay for the ease of Pavlovian conditioning is that you don't actually have any control over how behavior will change. This was a challenge in our case, since we knew exactly what set of behavioral changes we wanted to see. We wanted dogs to stop leaning on the walls, to cease charging around and jumping in their kennel, and to start paying polite attention to a visitor.

If you give this remit to a professional animal behaviorist, she will instigate a program of training that involves careful observation and timely rewards whenever the appropriate behavior occurs. While we knew that most shelters would not be able to afford this sort of training, we wanted to see how our cheaper, Pavlovian alternative would stack up against it, so we conceived of a study that would assess both methods.

Our study evaluated a group of dogs trained with rewards, the way the best professionals do it, alongside a group of dogs to which we applied Pavlovian techniques. To train the Pavlovian group, Sasha and her helpers walked up and down the shelter, ringing a bell and throwing in food treats. Later we compared the responses of the group that had been reward-trained to those of the Pavlovian group (and a control group, which heard the bell but experienced nothing else) when a stranger walked up to each dog's kennel. We found that both the reward-trained group and the Pavlovian group showed massive improvements in their response to a visitor. It looked like the reward-trained group had a small edge over the Pavlovian group — but that difference was tiny. The reward-trained group's behavior improved because we in effect "paid" the dogs with food treats for better behavior. Exactly why the Pavlovian group's behavior improved is more mysterious. Possibly, the expectation of imminent food just led to more of the friendly, attentive behavior that we know adopters like. Ultimately, we don't need to care why the behavior of the dogs in the Pavlovian group improved — all that matters is that it did. The crucial take-home from this test was that both experimental groups improved enormously, compared to the control group that just heard the bell.

Here we had what we had set out to find: a procedure readily applicable to large numbers of dogs by anyone inclined to do it, whether or not they had any expertise in training animals. Aside from the risk of tripping over something, the Pavlovian procedure of ringing a bell and throwing some food into the dogs' kennels could be applied by someone with her eyes closed. The only thing that made the procedure at all cumbersome was my—admittedly rather childish—imposition of the bell. It just appealed to my sense of humor to have a student born in Russia carry out an experiment on dogs using Pavlovian conditioning, with a bell as the conditioned stimulus. The fact that the fabled bell sprung from a mistranslation of the original Russian did not dampen my enthusiasm.

Our follow-up studies have shown that the bell is quite unnecessary: the presence of a human being alone can act as a conditioned stimulus. This means that a shelter needs to do no more than assign people to occasionally walk around the kennels, throwing in food treats. These people don't even have to be on the staff of the shelter; they can be visitors looking for a new dog. This technique will improve the dogs' behavior in a way that makes them more adoptable. Just giving treats to dogs is a great way to improve their behavior, and it helps them find new homes. This method, which demands almost nothing of the shelter and its hard-pressed personnel, inhibits the problem behaviors that develop when a dog is locked in a kennel for more than twenty-three hours a day. It also helps dogs communicate how positively they feel toward people. It opens out their natural desire to project love toward people and thus helps them find a place in a new human home.

I am very proud of the studies that Sasha carried out while we were working together, showing that interventions at the shelter that require no real expertise can improve dogs' chances at getting adopted. A more recent collaborator, Lisa Gunter—once my PhD student and now my colleague at Arizona State University—has actually figured out a way to increase adoptions while *reducing* a shelter's workload. By making a simple change in the way that shelter dogs are identified, we can ensure that more dogs will get a chance to trade on their loving nature when negotiating for a permanent place in a human household, just as their ancient ancestors did.

Before Lisa and I started working together, she had many years' experience in shelters in different parts of the United States. She was struck by how many people who came in looking for a dog to take home didn't really pay much attention to the dogs themselves. A lot of people had a fixed idea that they wanted a certain breed of dog; as a result, they ignored dogs whose kennels weren't labeled with that breed name.

This struck Lisa as odd for several reasons. For one thing, most of the dogs in shelters are mutts — animals of mixed parentage. The breed labels that shelters stick on kennels are nothing but guesses. In research that Lisa and I carried out together, we found that around 90 percent of these guesses are wrong. Whereas it is widely believed that about a quarter of shelter dogs are purebred and the rest are mixtures of two breeds, we discovered that barely one dog in twenty is a purebred, and the remainder contain DNA signatures from an average of three breeds — and sometimes as many as five. By using a painless mouth swab to get dogs' DNA and carrying out basic genetic tests on these samples, we had shown that breed labeling is even more of a charade than we had thought.

To be fair to shelter staffers, there are more than two hundred breeds registered today, which makes the task of guessing a dog's breed background exceedingly difficult. It is even harder because genes do not act like paint colors: when you mix genetic backgrounds, the result is not some kind of simple compromise between the races of dogs you started with, the way that mixing red and yellow makes orange. Rather, it is the outcome of a very high-level interaction, such that the offspring may look much more like one parent than the other — or quite commonly, not much like either parent.

Given the magnitude of the task of assigning breeds to dogs in shelters, it is hardly surprising that shelter staff don't often get them right. But the sad thing is, people are more swayed by the breed label itself than by the dog it describes — the animal that is right in front of them, demonstrating its companionable, loving nature with every wag of its tail.

Lisa decided to test the power of these inaccurate breed labels to guide the decisions of would-be adopters. She did this by focusing on one highly charged label in particular: that of the pit bull.

Despite what you may have heard, the pit bull is not actually a breed of dog. Rather, it is a label commonly applied to dogs of certain stocky builds, particularly dogs that look at least a bit like various terrier and bulldog breeds, such as the American Staffordshire terrier and the American bulldog. As Bronwen Dickey explains in her thoroughly grounded and fascinating book *Pit Bull: The Battle over an American Icon,* these dogs became pariahs in the late twentieth century through a complex convergence of cultural factors, which have nothing at all to do with the personalities of the dogs to whom this label is applied. I'd be inclined to say pit bull is just a blanket category used to tarnish the reputation of dogs with a certain characteristic body form. And Lisa's research clearly shows that the label of pit bull is applied so inconsistently that there is no strict characterization of what one looks like, never mind how one acts.

Knowing that labeling a dog as a pit bull would be a trigger for many potential adopters, Lisa devised an elegant experiment in which she leveraged the term's fearsome connotations. She put together one set of pictures and videos of dogs that had been labeled as pit bulls, which were then in residence at a shelter in Arizona. Another set of images of dogs in the same shelter looked identical to one of the pit bull dogs, but, for whatever reason, they had been labeled as a different breed.

The fact that these dogs had managed to avoid being labeled as pit bulls is a little remarkable. If you haven't spent a lot of time looking at dogs available for adoption at shelters in the United States, you might be surprised by just how wide a range of dogs is assigned this label: so-called pit bulls range from black to fawn in color and from small to medium in size; some have the blocky head I think of as characteristic of the pit bull, while others have a more elongated snout, like a retriever's.

The looseness of this definition now worked in Lisa's favor. She was able to amass an interesting collection of diverse dogs, which she organized in pairs of similar-looking beasts. In each pair, one had been labeled as a pit bull by the shelter, and the other was a "look-alike" that had somehow escaped that moniker, despite looking so much like a pit bull.

When Lisa showed potential adopters the photos and videos of these dogs without any breed labels—without any information at all, really, beyond the images on a computer screen—she discovered something surprising. Her subjects found the dogs that the shelter had labeled as pit bulls to be slightly more attractive and adoptable, on average, than the ones that had received a diverse array of alternative breed labels. When Lisa repeated the study, this time including the name of the breed that the shelter had pinned on each dog, the pit bulls' attractiveness plummeted.

Having spent less time hanging around shelters than Lisa had, I was more surprised than she was by this finding—that the name "pit bull" could influence people's judgment even more than the way a dog looked or acted. We were both very disappointed. Breed labels at shelters are wild guesses and very unlikely to accurately capture the breed heritage of the dogs, yet they are a stronger determinant of a dog's fate than anything the animal can do. Dogs' loving personalities, it seemed, were no match for an empty, arbitrarily assigned breed label.

But this sad finding gave Lisa an interesting idea. *What would happen,* she wondered, *if shelters gave up trying to guess the breeds of their dogs?* We agreed that, by dispensing with breed labels, shelters could probably help the dogs that might get labeled as pit bulls; after all, Lisa's research had showed that prospective adopters liked these dogs plenty if they saw them without the objectionable label. But what would getting rid of labels do to the dogs that might have been labeled with breeds that people actually like, such as spaniels and golden retrievers? Would we in effect be playing a shell game, simply reassigning happy and sad outcomes to different dogs? Or would we be helping all dogs across the board?

Lisa and I were discussing the pros and cons of this idea, and strategizing about how to find a shelter that might be willing to conduct a trial for us, when we heard that—by a beautiful coincidence—a major shelter in Florida had already done exactly what we were looking for. On February 6, 2014, Orange County Animal Services, a large animal shelter run by the local government in Orlando, Florida, stopped putting breed information (or rather, breed guesses) on their dogs' kennel cards. Very kindly, they provided us access to their intake and outcome data for the twelve months before they made the big change, and the twelve months afterward. Lisa and our collaborator Rebecca Barber crunched the numbers for over seventeen thousand dogs.

The results were very cheering. As we expected, dogs that might have been branded as pit bulls fared far better without that damning label; their adoptions increased by 30 percent. But the even better news was that adoptions for *all* breed groups increased.

There were no losers under this new setup. Even dogs categorized as toy breeds—usually among the most easily adopted at any shelter—showed a small improvement in adoption rates, and no group of dogs experienced reduced adoptions.

Subsequently, Orange County Animal Services let us see their data for a second year, during which they continued to omit breed "information" on dogs' kennel cards. Adoptions of all dogs continue to be higher than they were before the breed labels were removed. We were thrilled to see that the trial's initial success had not been some flash in the pan. It had yielded real improvements in outcomes for all dogs. And as a bonus, it had actually reduced the duties of the shelter staff, saving them from wasting their time in guessing at the breeds of all the dogs in their care.

Lisa and I have pondered why the outcomes for all dogs improved when the breed information was removed. We'd expected that getting dogs out from under the "pit bull" label would help them. But we had been puzzled to see that this change appeared to help all the dogs get adopted—including those that might have been sitting behind kennel cards with perfectly attractive breed names written on them. We discussed this at length, and the best hypothesis we have been able to come up with is the following:

When people visit a shelter, looking for a new dog, they define for themselves the kind of dog they want mostly by naming a breed. During childhood they may have spent many happy hours with a delightful German shorthaired pointer. So, now that they are parents themselves, looking to produce similar happy memories for their children, they show up at the shelter and ask to see a German shorthaired pointer.

To understand what happens next in this hypothetical scenario, we need to keep in mind these three facts. First, German shorthaired pointers are not especially common in the United States. Second, most of the dogs at a shelter are mixed breeds. Third, none of the dogs have papers. A shelter might have a hundred dogs available for adoption. Many may have the body size (medium to large) and the energetic, playful, tolerant, and affectionate nature that our hypothetical shelter visitors need for their family. But the odds are very slim that the person at the shelter responsible for dreaming up the breed names that identify the dogs will have thought of putting "German shorthaired pointer" on any of the kennel cards.

And so this imaginary couple go home empty-handed. They may not even look at the dogs if the office says that no German shorthaired pointers are up for adoption. These would-be adopters may consider the search aborted before it has even begun.

Now consider what happens when someone arrives at a shelter and is told there is no breed information available for any of the facility's dogs. This visitor may at least actually go and look at the dogs for himself. He may see a dog that reminds him, just because of its demeanor, of the dog of his childhood. Or, if the kids have come along, they may identify a dog that speaks to them of future adventures.

Taking away the breed labels actually frees people to look at the dog in front of them. Many dogs, of widely diverse body forms, can give people what they seek in canine companionship: a beer buddy, perhaps; someone to watch TV with; a hiking pal; love.

Whatever the reasons behind them, the study's findings make one thing clear: by looking beyond labels, we can help thousands of dogs, perhaps millions, to find homes. Truthfully, I think we should look past the breed in *all* our dealings with dogs. Aside from certain breed-specific behaviors (such as herding and pointing) that are not important to most people nowadays, a dog's breed information doesn't tell much about the animal's personality. This was demonstrated by two large studies that investigated the personalities of many thousands of purebred dogs from a large number of breeds. The researchers found that the differences in personality between dogs of the same breed were as large as — or, in some cases, even larger than — the differences in personality between dogs of different breeds. This finding actually seems fairly unsurprising when you consider that, as I outlined in the previous chapter, even cloned dogs — animals whose DNA is identical — do not necessarily have similar personalities. Why should dogs that are simply descended from the same ancestors, and therefore have much greater variation in their genomes, be any different?

Humans have a lot to gain by keeping an open mind toward breed labels. This point was driven home by one of the other findings of Sasha's research: people visiting a shelter typically ask to look at only one dog. They either go home with that dog, or they go home with no dog at all. It would serve dogs and people better if we went to the shelter looking for a dog without relying on arbitrary and arguably irrelevant markers such as breed labels. Now that many shelters encourage prospective adopters (and others) to foster dogs, it's easy enough to take a dog home for a weekend's trial run, to see how he fits into your family outside the stressful shelter environment, regardless of his "breed." If you take home a dog that perhaps doesn't look like the dogs you have known in the past, you might be surprised by the love he has in him.

And that is the essential thing. Few of us need a dog to do anything in particular. We just want a loving companion, and dogs deserve a fair chance to show us how they can fill that role. Give them that opportunity, and they'll prove that love can be found in dogs of any breed or no particular breed at all.

The shelter is a dog's last hope, and dogs in shelters are charity cases. So it is tragic but not entirely surprising that they are not treated more generously.

I'm much more surprised that, at the other end of the economic spectrum, dogs' lives can be as poor as many of them are. For it's not just mutts at the shelter that need our help; the purebred aristocracy of the dog world deserves better too. These dogs are just as capable and deserving of forming supportive and loving bonds with people. They too are signatories to the ancient pact between our species; they too deserve to lead fulfilling lives. Yet too many purebred dogs are in as much of a crisis as many shelter dogs are, albeit in a quite different way.

Dog breeds, as we understand them today, are a product just of the past 150 years. Detailed DNA analysis has demonstrated that breeds spoken of as ancient are actually no older than the past century or two. Even the saluki, which looks like the noble hounds painted on the tombs of Egyptian pharaohs thousands of years ago, was created as a breed in the modern sense back in the nineteenth century. Earlier than that, people certainly recognized distinct general shapes of dogs; artworks from ancient Egypt, for example, suggest the existence of four or five recognizable distinct forms of dogs, and Roman literature names forty or fifty races of dogs. But these were not "breeds" as we understand that concept today. That is, they were not intensely isolated populations, completely closed off from the possibility of breeding with dogs of any other group — with all of the genetic perils that such sequestration entails.

Many people are unaware of just how intensively inbred these "pure" dogs have become. If you look at a purebred dog's family tree, it is not uncommon to see that his father is also his grandfather and his mother's uncle to boot. This intensive inbreeding ensures that purebred pups inherit their looks (if not their personalities) pretty reliably, but it also brings with it serious risks. Purebred dogs have shorter life expectancies than their mixed-breed cousins, for instance; this is because they tend to suffer from a wider range of health problems than dogs with more diverse backgrounds experience.

Colleagues of mine at Arizona State University's Biodesign Institute, Carlo Maley and Marc Tollis, together with their student Cassandra Balsley, recently completed a thorough analysis of the causes of death in over 180,000 dogs from over two hundred distinct breeds from around the world. They found that among some breeds of dog, more than half of all individuals die from cancers; the more inbred the dog breed, the higher the rate of cancer deaths.

Carlo and Marc explained to me that, when people started creating modern dog breeds in the nineteenth century, they had enough of an understanding of inheritance to know that breeding together closely related animals with traits they liked the look of would increase the chance that the puppies would share those traits. What dog breeders didn't know then—something widely understood today—is that, as obvious and desirable traits are captured into the animals' genes by this process of inbreeding, undesirable and hidden traits also get captured. Consequently, many races of purebred dogs show shockingly high rates of cancer—the archetypical hidden genetic evil—as well as other inherited diseases. Dalmatians are predisposed to deafness, boxers to heart disease, and German shepherds to hip dysplasia—to give just three examples from a depressingly long list of maladies.

Happily, the plight of purebred dogs has of late been receiving more and more attention. The United Kingdom Kennel Club—the granddaddy of all breed clubs around the world—was deeply shamed a decade ago by a documentary on BBC Television, *Pedigree Dogs Exposed*. This program, which went on to win the highest accolade for animal welfare from the Royal Society for the Prevention of Cruelty to Animals, drew attention to the practices of intense inbreeding and their consequences for animal welfare.

The Kennel Club came out of this documentary looking very foolish. For example, confronted on the ethics of mother-to-son matings, the chairman of the club at the time, Ronnie Irving, commented that it "depended on the individual mother and son," and added, "I don't want a bunch of scientists telling me they know more about it." One of those scientists, Steve Jones, a world-renowned geneticist at my alma mater, University College London, summed up the bleak outlook for purebred dogs: "If the dog breeders insist on going further down that road, I can say with confidence, really, that there is a universe of suffering waiting for many of these breeds—and many, if not most, of these breeds will not survive."

The BBC documentary motivated Parliament to call for an independent inquiry into the welfare of purebred dogs. It was led by Professor Sir Patrick Bateson, Fellow of the Royal Society and widely considered one of the world's leading behavioral biologists. He concluded that, although many people in Britain bred dogs with the utmost concern for their welfare, the business of inbreeding for purebred dogs was out of control. He cited a study from Imperial College London, which found that, although there were nearly twenty thousand boxers in the UK, these animals carried the genetic equivalent of just seventy distinct individuals. Britain's ten-thousand-plus pugs were genetically equivalent to a population of fifty individuals.

I can understand how people prefer the look of some dogs over others; I do too. And I can understand why breeds of dog exist: some people want long golden hair, some want short curly white fur, some people want long, wolfy snouts, and some want dogs with short faces. None of that is hard to comprehend.

What I just cannot understand is the obsession with knowing that the genes in your dog come from only a small set of dogs that were selected, during the Victorian era, to be the founders of a breed. Why is it important to owners of German shepherds to know that their dog, in 2019, can follow its lineage back to one of the dogs that Rittmeister (Cavalry Officer) Max Emil Friedrich von Stephanitz decided, back in the late nineteenth century, was a perfect exemplar of the dogs kept by shepherds in Germany? This to me is a mystery — and a deeply worrying one, at that.

Many of the problems of purebred dogs today could be corrected by allowing the interbreeding of small numbers of dogs from related breeds. This limited outbreeding would have minimal impact on the dogs' appearance but could tremendously improve their health. If we are to go about righting the systemic wrong that is the purebreeding of dogs today, and to reciprocate their love by giving them the fond care we owe them, this would be a massive step in the right direction — one that would require minimum concessions from breed enthusiasts.

For example: Every registered dalmatian in Britain suffers from a genetic defect called hyperuricosuria, which impacts its ability to metabolize uric acid. As a consequence, these dogs can suffer a variety of difficulties, many painful, which ultimately lead to an early death. Back in the 1970s an American geneticist and dog breeder, Dr. Robert Schaible, started breeding pointers with dalmatians to correct this defect. So far as the genetics of this uric acid problem was concerned, Schaible's program was a complete success, and anyone looking at the dogs he created would see what appear to be beautiful dalmatians. When one of his dogs, Fiona, a fifteenth-generation descendent of the original dalmatian-pointer cross, 99.98 percent genetically pure dalmatian, was brought over to the UK to compete in the Kennel Club's major competition, Crufts, local breeders were up in arms: "It is pretty unethical to allow this dog in a pedigree show. As far as I'm concerned it is an illegal entrant and makes a mockery of the Dalmatian breed," said one breeder. Another concurred: "This is a mongrel. This is unethical and I'd be disgusted if the dog won." I have to wonder how these people define "unethical." To be clear, to anyone who hasn't seen the dogs' pedigrees, the differences between Schaible's dalmatians and the "pure" variety are so imperceptible as to be nonexistent. The British newspaper *Daily*

Mail printed a photograph of Fiona alongside a "regular" dalmatian. Nobody would ever be able to tell them apart by looking at them; the only distinction lies hidden in their genes.

This story about dog breeds, at least, has a happy ending. Fiona did not win at Crufts, but she did win the right to be registered with the Kennel Club as a dalmatian, so that her healthy genes can be bred into the dogs of Britain and help produce healthy dalmatians — at least among those breeders willing to tolerate her 0.02 percent racial impurity.

The underlying problem here, to my mind, is that some people value the purity of a dog's bloodline more than its capacity — indeed, its eagerness — to have a loving relationship. Can a dog's pedigree really be more important than that?

Shelter dogs and purebred dogs alike are defenseless before another type of human depredation: lax governmental regulations, which tolerate human action toward dogs that are far from giving dogs the lives they need and surely deserve. This is an issue in many parts of the world, but because I live and work in the United States, I have seen the problems with the American regulatory system firsthand and in depth.

Because my students and I carry out studies on people's dogs, my employer, Arizona State University, very properly demands that I read and obey the federal law regarding animals: the Animal Welfare Act. If you are based in the United States, you should read this law. I think you will be as shocked as I was.

The Animal Welfare Act is the federal law that regulates the behavior of dog breeders or anyone else in the business of making a living off animals. What stands out when you read this law is that it makes no attempt to define "animal welfare." The purpose of the act, it tells us, is to regulate commerce in animals — not, strange as it may sound, to promulgate their welfare.

Among many other things, this law lays down how dogs in breeding establishments may legally be held in the United States. The standards are shockingly out of touch with dogs' needs and the expectations of people who buy these dogs. To take one depressing fact from a vast pool of sad failures: this law decrees that a cage just six inches longer than the dog (not even counting its tail) is acceptable for an animal's lifelong housing. If the cage is just twice that pathetically inadequate size, the regulations do not require that the poor animal ever leave its cage at all, not even for an hour of sunlight, and certainly not to form relationships with other beings.

Xephos is about thirty inches long, from her snout to the base of her tail. Consequently, the law permits her to be held in a cage measuring just thirty-six inches on each side. She wouldn't even be able to wag her tail in that kind of space (though I doubt, forced to live in such a tiny cage, that she would be inclined to wag her tail). To make an image to accompany a talk I gave about this issue, I marked out a thirty-six-inch square on the ground and asked Xephos to sit in it, so I could take a photo. She looked so miserable and confused that I would ask her to do this even for a few moments—imagine the dogs that live out their whole lives like this! It beggars belief that a law titled the Animal *Welfare* Act could in this and so many other respects pay no attention to the needs of animals.

The inadequacy of current legal protections for animals in general, and our loving canine companions in particular, has been receiving more and more attention in recent years. For instance, the journalist Rory Kress explores the tragic truth about dog breeding in the United States in a sad but wonderful 2018 book called *The Doggie in the Window*. She doesn't go after the illegal backyard operations that people call "puppy mills," but instead concentrates on the inhumanity that the law tolerates in regulated facilities. Kress tells a personal story of trying to uncover the origins of a puppy she bought on a whim from a pet store. I won't give away the ending, but suffice it to say that she journeys down a rabbit hole of inadequate regulation and callousness.

As dog lovers, and people who understand how dogs love us and the responsibilities that this puts on us, we should not be tolerating such weak protections for our canine fellow-travelers. Of all the ways that we can make our dogs' lives better and honor the love they heap upon us, fixing these inhumane regulations may be the most difficult. But it also stands to have the greatest impact on dogs' well-being: not just those with whom we share our homes, but those with whom we share our country. As informed citizens, we should be demanding nothing less.

Humans have repaid dogs' love with some terrible mistreatment, but nevertheless, I remain optimistic about people and dogs.

One of the things that makes me optimistic is that I know dogs are resilient. I mentioned earlier that sweet Xephos had led a difficult life before we adopted her and that she has recovered, with no apparent ill effects. This speaks to an uplifting fact: dogs can be rehomed very happily. They do not appear, as our species does, to suffer lasting trauma at losing an important attachment figure. This is likely because, among their own kind, dogs do not seem to form the same lifelong bonds that we do.

Studies my students and I have carried out — as well as much everyday experience around dogs — show that these animals are more flexible in their relationships than we are. We have seen that dogs start to form new bonds in a matter of minutes, and even street dogs quickly affiliate with people who treat them gently. This is not to say that dogs don't remember the people they love; they certainly do. Charles Darwin, when he came back from voyaging around the world on the *Beagle* for five years, was struck that his dog at home still remembered him, and Xephos tells me, with an almost embarrassing intensity, how she has missed me when I return to her after any time away. But it is valuable to know that dogs can recover from earlier traumas — that they are resilient. (One implication of this: there is no reason to hesitate in adopting an older dog out of concern that it might pine forever for the family it has lost. But it should go without saying that dogs' resilience is no excuse for abusing them or depriving them of important emotional bonds unless that really cannot be helped.)

The other reason I am optimistic that we will do better by our dogs is because so many people are determined to do just that. Wherever I go, I meet people who reciprocate in full measure the love their dogs express toward them. I see this from the wealthiest people I meet in the United States, whose pedigree pooches enjoy soft beds and expensive diets, through to the homeless people sheltering under bridges who share what little they have with the dogs that offer them loving support during difficult times. Wherever I travel, I find people caring for dogs — be they the street dogs of Moscow, which get sustenance from busy commuters outside subway stations and take shelter in cardboard boxes that apartment dwellers put outside to protect them from the snow; or the pets of Tel Aviv, which get exercise at one of that city's many lovely dog parks; or the dogs in Nicaragua, whose Mayangnan masters, in an unfussy way, keep their dogs close and do what they can to keep them healthy.

People *love* dogs. If that verb means half as much to us as it does to them, we will do the hard work necessary to give them better lives, and to honor all that they give us. Dogs' love defines them. Theirs is an example we should follow.

CONCLUSION

IF THIS JOURNEY has changed you the way it has me, you will go forward more attuned to—and appreciative of—dogs' love than you were before.

The littlest habits of our dogs remind us that they love us. On a normal day, when I'm at my desk at home, Xephos curls up at my feet or on the carpet just behind me; if I'm reading in bed, she lies down at the foot of the bed, placing her back against my feet; if I'm slow in clearing up after dinner, or go back to my desk, she'll start warming her place on the sofa in expectation of the TV watching to follow. If a visitor doesn't understand that Xephos wants to be petted, she will push herself under that person's hand, a prompt to stroke her head.

Such behaviors will be heartwarmingly familiar to many dog lovers, and they take on new, powerful meaning when you appreciate the fascinating science and rich history behind these demonstrations of affection.

These beautiful hallmarks of our lives together are all the more poignant when you consider that dogs' expressions of love so often go unreciprocated. For instance, Xephos loves snuggling up to people when they're in bed. My wife, Ros, and I always let her sleep at the foot of our bed, with us; once or twice, we've even let her crawl under the covers to snuggle. We once had housesitters who, not unreasonably, did not want to follow our established practice of letting Xephos onto the bed. Poor Xeph' cried and cried; finally, when she realized she wouldn't be allowed up, she crept under the bed to sleep.

Xephos is a resilient creature, and she quickly bounced back. Still, whenever I think of her plaintive, rebuffed attempt to express her love, I feel sympathy for her affectionate confusion. It is a reminder that dogs' love isn't projected out into a vacuum. Those of us who enter into a relationship with them (even if just a casual housesitting arrangement) owe it to dogs to hear and honor these expressions of their emotional needs. If we don't, we may inadvertently cause these animals real suffering.

I believe this now with every fiber of my being, but of course I once was skeptical of the idea that dogs express love in their interactions with humans—or even that they have any love to give us at all. So I probably should not be so dismayed when I am confronted with this same skepticism from some of the people with whom I share this theory of dogs' love. And I do regularly encounter doubters, some of whom are quite adamant that the very notion of dogs' love is rubbish.

Early in my quest to understand how dogs love people, I was sitting next to a random person on a plane in whom I unwisely confided my developing belief in what makes dogs special. Not only was he adamant that dogs do not care about people, but I also had to discourage him from showing me a substantial scar he said he had on his thigh, from the time he tried to separate two fighting dogs.

Dogs' behavior is certainly not confined to happy, loving smiles and tail wags, and it is indisputable that dogs do sometimes harm people. In the United States, there is no definitive record of how often dogs bite people, but how much money gets spent due to this problem is logged fairly reliably. US insurers paid out $686 million for dog bites in 2017—an astonishing amount of money. The large figure, however, stems from the amount paid per claim (over $37,000) rather than the number of claims (18,500). In a nation that contains around eighty million dogs, 18,500 claims is not a big number. To put it into context, it means each dog in the country will bite somebody badly enough to prompt an insurance claim roughly once every five centuries — and thankfully, dogs don't live to be anywhere near that old. To be sure, these numbers could be out by a considerable margin because of people who get bitten but don't have any insurance, and others who are bitten but can't find anyone to sue, but even so it is pretty clear that dogs are not a great danger to people. The vast majority of dogs lead peaceful, harmless lives.

In any case, the fact that a loving relationship may exist between members of two species in no way rules out the possibility that individuals of those species also may cause injury to each other. People can form loving relationships with other humans, but people nonetheless do each other a great deal of harm. Human-on-human violence in the United States costs over a thousand times more than dog-on-human violence. If the eighty million dogs in the United States were people, they would kill around four thousand other people each year. Since they are only dogs, they are responsible for fewer than forty deaths each year. You are vastly safer in the company of a dog than with another of your own species.

Sadly, and in much the same way, just because dogs are capable of loving humans does not mean they *always* love us. And just because they appear to show clear signs of affection, that does not mean that their behavior cannot also reflect at times other deep-seated, powerful emotions, such as fear or anger.

It is also spurious to suggest — as did a friend I will not name here — that dogs' seemingly loving behaviors are expressed not out of love, but rather self-interest: that is, dogs make us think they love us in order to trick us into caring for them. Of course, appearing to love us *would* be in dogs' interest; after all, many of us care for and love our dogs in large part because we perceive our affection is reciprocated — for the simple reason that love begets love. And I suppose that, if I tried hard enough, I could imagine an animal that could wag its tail enthusiastically without being happy, or seek me out without really caring about me. I find that tough to envisage, but not entirely impossible. But what about all the physiological evidence I've uncovered in recent years? From the genes that code for loving behavior, through the brain states that register and direct dogs' affection for people, the hormones that match the activity found in our own species when we feel love toward other individuals . . . ? We cannot throw away all the powerful evidence that love is a real thing in dogs' lives. The weight of scientific evidence is, I am convinced, too great for skepticism about dogs' love to be a viable position to maintain any longer. And I say that as someone who, I think, held out longer than most people would against the idea that dogs could love us.

But just suppose, despite all the evidence I've presented in these pages, that your dog really does not love you — that he or she is just faking an affectionate reaction to your presence. Now look at your spouse. Could she or he be faking affection toward you? And how about your child? Your best friend?

The truth is, there is no absolutely surefire, completely unimpeachable way of knowing that anyone in your life who appears to care for you really feels love for you. Over time, we build up a nuanced sense of how different people in our lives feel toward us, on the basis of our experiences with them — experiences in which their comportment and behavior reveal an enormous amount about them and how they feel toward us. I can see no reason to be more skeptical toward our dogs than we are toward anyone else who appears to love us. If anyone loves you, your dog does.

It took me a long time to come to this conclusion, and the experience has fundamentally changed the way that I relate to dogs. Yet of all the lessons I have learned on my journey with Xephos, the wolves at Wolf Park, and the many other canids that have helped us in our studies, one moral stands above the rest. It is a lesson I take with me into my interactions not only with our canine companions, but also with my fellow humans.

There is a movement in our culture today that equates strength, especially but not only manly strength, with exploiting whatever power one has over others—be it physical strength, elite social standing, or financial prowess—at the expense of those who are weaker. This surely is a brutal morality—a "dog eat dog" attitude to life that ill becomes people or their canine friends.

But there is another concept of strength, and that is the power to help the weak—to support those who are less able to fend for themselves. I'm not a religious person, but I honor and respect the great spiritual leaders who over the millennia have taught that we find our greatest strength when we aid the weakest among us.

Recognizing and freely reciprocating the call that dogs make on our affection is a way of practicing that second form of strength. In loving dogs as they love us, we tap into and reinforce our best, most altruistic selves. There is honor and decency in this selflessness, and our relationships with dogs and humans alike are elevated when we practice it.

Of course, dogs do reciprocate the support we give them in many diverse ways, from the guarding functions of the most ancient dogs on prehistoric trash dumps, through the dogs that helped hunters and assisted our ancestors during a very difficult period in human evolution at the end of the last ice age, up through the modern dogs who perform a whole range of ingenious support roles after extensive training. Also, increasing numbers of research studies indicate that people who have dogs lead healthier, happier lives than those who do not.

I'm a little skeptical of those studies (have I mentioned I tend to be a little skeptical?). I certainly think I'm happier having Xephos as part of my life than I was during the years my home did not harbor a dog. But I think it might well be that, on average, people who choose to add a dog to their home are already healthier and happier than people who cannot find space in their lives for canine companionship. In my own case, it was an uptick in stability that made it possible for Sam, Ros, and I to invite a dog to join our family.

Whichever way the evidence falls out on this question, I don't believe we should care for dogs simply because they are useful to us. I don't like to think of human-canine relationships as transactional. If it was, caring for my dog would be similar to taking care of my car. My car is a necessary part of my life; it performs certain useful functions, and consequently I do what I have to do to keep it running smoothly. But a dog does so much more than complete a set of functions. It can call upon wells of affection that we may not know we possess, and encourage us to selfless action in response to another living creature. A dog can surprise us, and make us surprise ourselves.

We should care for our dogs because they deserve it. We show true nobility when we respond to our dogs' pleas for support without considering whether there is any way they might pay us back. When we reach out in this way, we are responding to the unspoken but nonetheless binding promise between our kind and theirs — a social contract that reaches back in time well beyond the day I first saw poor little frightened Xeph' cowering in a kennel at a noisy shelter, all the way to however many millennia ago her kind first acquired the genes that make possible her extraordinary capacity for love. When I respond to Xephos, I am following in the footsteps of millions of people over countless centuries: not just Pavlov, Darwin, and Arrian of Nicomedia, but also whoever first noticed the mute plea for support from a pup somewhere near a human village hundreds or (more likely) thousands of years before that, and answered that cry for help — causing the dog to imprint on him or her, and cementing a bond that has linked our two species ever since.

These humans and their dogs have been participants in a cross-species partnership that spans the ages. It is a wonder, and an honor, to participate in it. To be loved by a dog is a great privilege, perhaps one of the finest in a human life. May we prove ourselves worthy of it.

ACKNOWLEDGMENTS

If one of the pleasures of reaching the end of a project like this is the opportunity to thank the many people who have supported me along the way, one of the anxieties is the possible failure to mention someone who did me some great favor. If you are that person—my profound apologies.

Like the leader of a band, let me start by calling out the fantastic soloists who have performed alongside me on this journey to get to the root of what makes dogs the amazing beings they are. In order of appearance: Monique Udell, Nicole Dorey, Erica Feuerbacher, Nathan Hall, Lindsay Mehrkam, Sasha (Alexandra) Protopopova, Lisa Gunter, Rachel Gilchrist, and Joshua Van Bourg—take a bow. In addition to these amazing graduate students, armies of undergraduates have also provided indispensable assistance to our work. I am grateful to every one of them, and I'm sorry there isn't space here to name you all. I also want to thank Anne-Marie Arnold, Mariana Bentosela, Nadine Chersini, Jessica Spencer, Robson Giglio, Kathryn Lord, David Smith, Maria Elena Miletto Petrazzini, and Isabela Zaine, for their diverse assistance with our research over the years. As Neil Young said when he went on stage with The Band at The Last Waltz in 1976, "It's one of the pleasures of my life to be on this stage with these people." I mean it.

In our studies we have received so much help from many people who struggle at animal shelters to give their charges the best lives they can with the resources they have — my gratitude to all of you for what you do for dogs, and for your patience with our research. At Wolf Park, Pat Goodman, Gale Motter, Monty Sloan, Dana Drenzek, Holly Jaycox, and Tom O'Dowd, as well as many other staff and volunteers, were infinitely indulgent of our demands on their time as we came up with ever crazier hoops to put their wolves through. Thank you for your patience and your friendship. To the hundreds of people who have let us test their beloved pets: thank you for your trust and your assistance.

On my travels I have often profited from the support of strangers who quickly became friends. Thank you to Jeremy Koster at the University of Cincinnati and his friends among the Mayangna; William Fielding at the College of the Bahamas; Ilya Volodin and Elena Volodina of Lomonosov Moscow State University and Moscow Zoo, respectively; Lyudmila Trut and Anastasiya Kharlamova of the Siberian Branch of the Russian Academy of Sciences as well as Anna Kukekova ("Russian, Anna, Russian! English, Anna, English!") at the University of Illinois Urbana-Champaign; Joseph Terkel and Eli Geffen at Tel Aviv University; Moshe Alpert and Yossi Weissler at Kibbutz Afikim; Ludwig Huber, Kurt Kotrschal, Sarah Marshall-Pescini, Friederike Range, and Zsófia Virányi at the Wolf Science Center in Ernstbrunn, Austria; Per Jensen at Linköping University and Hans Temrin at Stockholm University in Sweden; and Alliston Reid, as well as the sorely missed John Pilley, of Wofford College, South Carolina.

My gratitude to Angela Perri at Durham University and Greger Larson at Oxford University for teaching me a little archeology; Gregory Berns likewise for opening my eyes to cognitive neuroscience. What I know about dogs and oxytocin was explained to me by Takefumi Kikusui at Azabu University, Therese Rehn at the Swedish University of Agricultural Sciences in Uppsala, and Mia Persson of Linköping University in Sweden. Thank you all — any confusions that remain in this work are of course entirely my own responsibility.

I have wanted to write a book that might reach an audience beyond just my professional peers for a long time. Thank you to Aaron Hoover and Bill Canon for your advice and encouragement during that lengthy period; thank you to Steven Beschloss for helping bring this project to fruition. I owe you all a drink.

A special drum roll and applause for agent extraordinaire Jane von Mehren and her colleagues at Aevitas Creative Management. My editor, Alex Littlefield, extracted what I meant to say from what I actually wrote and, along with his enablers at Houghton Mifflin Harcourt, turned my thoughts into this tangible object you are holding in your hands — thank you, guys. My thanks too to Susanna Brougham for polishing the text into the shiny form you see before you and to Leah Davies for the sketches that enliven these pages.

By sheer good fortune I happen to work at a quite exceptional institution. Arizona State University somehow squares so many circles. I'm not going to repeat our advertising copy here, but believe me, it is an amazing place. Reputations are always lagging behind indicators. Maybe in fifty years' time, the world will have caught on to how brilliant ASU is now — a place where scholarship flourishes, where we pride ourselves on the people we give opportunities to, rather than those we exclude. (I wasn't able to entirely resist the advertising copy!) I am especially grateful to my colleagues in the Psychology Department for tolerating a "dog guy." I can't name you all here, so I'm going to name-check the two who have chaired the department during my time here — Keith Crnic and Steve Neuberg. Thank you for your friendship and for fostering an environment where we find strength in our gentleness.

The late Ray Coppinger, I fear, would have hated this book, but his mentorship looms over it, and I owe him a tremendous debt. I wish we could have the arguments I know this volume would have prompted.

My parents cannot escape some responsibility: both through phylogeny and ontogeny — nature and nurture — their influence is clear.

I suspect my father would have wanted to argue with me about this "dog love stuff" too.

Ros and Sam: what can I say? Thank you for sustaining me through all of life's ups and downs. Thank you for making it fun.

And Xephos too — this book's spirit animal, if ever there was one. I quite literally couldn't have done this without you. Thank you, sweetie. There's liver for dinner!

NOTES

Introduction

page

5 *genius of dogs:* Brian Hare and Vanessa Woods, *The Genius of Dogs* (New York: Dutton, 2013).

Chapter 1: Xephos

18 *a really quick learner:* Kathryn Bonney and Clive Wynne, "Configural Learning in Two Species of Marsupial," *Journal of Comparative Psychology* 117 (2003): 188–99.

as I read Hare's research: Brian Hare, Michelle Brown, Christina Williamson, and Michael Tomasello, "The Domestication of Social Cognition in Dogs," *Science* 298 (2002): 1634–36.

27 *"like wild animals":* John Paul Scott and John L. Fuller, *Genetics and the Social Behavior of the Dog* (Chicago: University of Chicago Press, 1965).

28 *"Debbie Downer":* Benoit Denizet-Lewis, *Travels with Casey* (New York: Simon & Schuster, 2014).

31 *But Monique and Nicole felt that:* M.A.R. Udell, N. R. Dorey, and C.D.L. Wynne, "The Performance of Stray Dogs (Canis familiaris) Living in a Shelter on Human-Guided Object-Choice Tasks," *Animal Behaviour* 79, no. 3 (2010): 717–25.

33 *"The world's smartest dog":* "The world's smartest dog, Chaser has the largest vocabulary of any nonhuman animal," Super Smart Animals, BBC Television, http://www.chasertheborderollie.com/.

34 *When I visited him and Chaser:* Author interviews with John Pilley, May 2009, Spartanburg, SC.

36 *John trained Chaser:* John W. Pilley and Hilary Hinzmann, *Chaser: Unlocking the Genius of the Dog Who Knows a Thousand Words* (New York: Houghton Mifflin Harcourt, 2013).

Chapter 2: What Makes Dogs Special?

50 *"Any policeman could tell you":* W. Horsley Gantt's preface to Ivan P. Pavlov, *Conditioned Reflexes and Psychiatry—Lectures on Conditioned Reflexes,* trans. W. H. Gantt (New York: International Publishers, 1941).

For eighty years: Daniel P. Todes, *Ivan Pavlov: A Russian Life in Science* (Oxford, UK: Oxford University Press, 2014).

52 *As a person came into:* W. H. Gantt et al., "Effect of Person," *Conditional Reflex: A Pavlovian Journal of Research & Therapy* 1, no. 1 (1966): 18–35.

53 *the same result over and over:* E. N. Feuerbacher and C.D.L. Wynne, "Relative Efficacy of Human Social Interaction and Food as Reinforcers for Domestic Dogs and Hand-Reared Wolves," *Journal of the Experimental Analysis of Behavior* 98, no. 1 (2012): 105–29. E. N. Feuerbacher and C.D.L. Wynne, "Shut Up and Pet Me! Domestic Dogs (Canis lupus familiaris) Prefer Petting to Vocal Praise in Concurrent and Single-Alternative Choice Procedures," *Behavioural Processes* 110 (2015): 47–59.

57 *Children whom Ainsworth termed:* M.D.S. Ainsworth, M. C. Blehar, E. Waters, and S. Wall, *Patterns of Attachment: A Psychological Study of the Strange Situation* (Hillsdale, NJ: Lawrence Erlbaum, 1978).

62 *Another Russian dog expert:* S. Sternthal, "Moscow's Stray Dogs," *Financial Times,* January 16, 2010, https://www.ft.com/content/628a8500-ff1c-11de-a677-00144feab49a. If you are interested in these dogs, there is a whole website dedicated to them, where Mus-

covites upload photos and videos of dogs they meet on their commutes—www.metrodog.ru.

64 *They can be a vector:* "India's Ongoing War Against Rabies," *Bulletin of the World Health Organization* 87, no. 12 (2009): 885–964.

Indian street dogs: D. Bhattacharjee et al., "Free-Ranging Dogs Show Age-Related Plasticity in Their Ability to Follow Human Pointing," *PLOS ONE* 12, no. 7 (2017): e0180643.

65 *"social reward is more effective":* D. Bhattacharjee et al., "Free-Ranging Dogs Prefer Petting over Food in Repeated Interactions with Unfamiliar Humans," *Journal of Experimental Biology* 220, no. 24 (2017): 4654–660.

Chapter 3: Dogs Care

74 *As Emma Townshend recounts:* Emma Townshend, *Darwin's Dogs: How Darwin's Pets Helped Form a World-Changing Theory of Evolution* (London: Frances Lincoln, 2009).

In one of his later: Charles Darwin, *The Expression of Emotions in Man and Animals* (London: John Murray, 1872).

"But man himself": Ibid., 11. One of 183 mentions of dogs in that book.

"with whom they were friends": Ibid., 119–20.

75 *"The upper lip":* Ibid., 122.

"It's as easy": Patricia McConnell, *For the Love of a Dog: Understanding Emotion in You and Your Best Friend* (New York: Ballantine Books, 2007).

78 *Recognition was slightly higher:* T. Bloom and H. Friedman, "Classifying Dogs' (Canis familiaris) Facial Expressions from Photographs," *Behavioural Processes* 96 (2013): 1–10.

80 *When left alone:* Dog's left and right. Reversed if looking at the dog head-on. A. Quaranta, M. Siniscalchi, and G. Vallortigara, "Asymmetric Tail-Wagging Responses by Dogs to Different Emotive Stimuli," *Current Biology* 17, no. 6 (2007): R199–R201.

82 *Bill was explaining:* K. MacPherson and W. A. Roberts, "Do Dogs (Canis familiaris) Seek Help in an Emergency?" *Journal of Comparative Psychology* 120, no. 2 (2006): 113–19.

83 *Several years later:* J. Bräuer, K. Schönefeld, and J. Call, "When Do Dogs Help Humans?" *Applied Animal Behaviour Science* 148, no. 1 (2013): 138–49.

84 *Ruffman and Morris-Trainor found:* T. Ruffman and Z. Morris-Trainor, "Do Dogs Understand Human Emotional Expressions?" *Journal of Veterinary Behavior: Clinical Applications and Research* 6, no. 1 (2011): 97–98.

In designing their study: D. Custance and J. Mayer, "Empathic-like Responding by Domestic Dogs (Canis familiaris) to Distress in Humans: An Exploratory Study," *Animal Cognition* 15, no. 5 (2012): 851–59.

88 *"It was Chum":* Louise Lind af Hageby, *Bombed animals . . . rescued animals . . . animals saved from destruction* (London: Animal Defense and Anti-Vivisection Society, 1941).

90 *The free rat:* I. B.-A. Bartal, J. Decety, and P. Mason, "Empathy and Pro-social Behavior in Rats," *Science* 334, no. 6061 (2011): 1427–430.

92 *"Dogs get lost":* Edward Thorndike, *Animal Intelligence: An Experimental Study of the Associative Processes in Animals* (New York: Macmillan, 1898).

Chapter 4: Body and Soul

99 *As Berns recounts:* Gregory S. Berns, *How Dogs Love Us: A Neuroscientist and His Adopted Dog Decode the Canine Brain* (Boston: New Harvest, 2013).

101 *Each of the two dogs:* G. S. Berns, A. M. Brooks, and M. Spivak, "Functional MRI in Awake Unrestrained dogs," *PLOS ONE* 7, no. 5 (2012): e38027.

102 *This activity in the centers:* G. S. Berns, A. M. Brooks, and M. Spivak, "Scent of the Familiar: An fMRI Study of Canine Brain Responses to Familiar and Unfamiliar Human and Dog Odors," *Behavioural Processes* 110 (2015): 37–46. P. F. Cook et al., "Awake Canine fMRI Predicts Dogs' Preference for Praise vs. Food," *Social Cognitive and Affective Neuroscience* 11, no. 12 (2016): 1853–862.

104 *"compact and energetic golden retriever":* Gregory S. Berns, *What*

It's Like to Be a Dog: And Other Adventures in Animal Neuroscience (New York: Basic Books, 2017).

105 *"we conclude that the vast majority":* C. Dreifus, "Gregory Berns Knows What Your Dog Is Thinking (It's Sweet)," *New York Times,* December 22, 2017, https://www.nytimes.com/2017/09/08/science/gregory-berns-dogs-brains.html.

106 *This substance was first:* W. Feldberg, revised by E. M. Tansey, "Dale, Sir Henry Hallett (1875–1968), physiologist and pharmacologist," in *Oxford Dictionary of National Biography*, rev. ed. (Oxford, UK: Oxford University Press, 2004), http://www.oxforddnb.com/view/10.1093/ref:odnb/9780198614128.001.0001/odnb-9780198614128-e-32694;jsessionid=A2331762884803A4CD2420C7D4200C59.

Vincent du Vigneaud: "Vincent du Vigneaud—Facts," NobelPrize.org (Nobel Media AB 2018), https://www.nobelprize.org/nobel_prizes/chemistry/laureates/1955/vigneaud-facts.html.

109 *The reason for this:* H. E. Ross and L. J. Young, "Oxytocin and the Neural Mechanisms Regulating Social Cognition and Affiliative Behavior," *Frontiers in Neuroendocrinology* 30, no. 4 (2009): 534–47.

110 *The research team at Azabu:* M. Nagasawa et al., "Dog's Gaze at Its Owner Increases Owner's Urinary Oxytocin During Social Interaction," *Hormones and Behavior* 55, no. 3 (2009): 434–41. S. Kim et al., "Maternal Oxytocin Response Predicts Mother-to-Infant Gaze," *Brain Research* 1580 (2014): 133–42. T. Romero et al., "Oxytocin Promotes Social Bonding in Dogs," *Proceedings of the National Academy of Sciences* 111, no. 25 (2014): 9085–90. M. Nagasawa et al., "Oxytocin-Gaze Positive Loop and the Coevolution of Human-Dog Bonds," *Science* 348, no. 6232 (2015): 333–36. T. Romero et al., "Intranasal Administration of Oxytocin Promotes Social Play in Domestic Dogs," *Communicative & Integrative Biology* 8, no. 3 (2015): e1017157.

113 *Dogs with the AA version:* M. E. Persson et al., "Intranasal Oxytocin and a Polymorphism in the Oxytocin Receptor Gene Are Associated with Human-Directed Social Behavior in Golden Retriever Dogs," *Hormones and Behavior* 95, Supplement C (2017): 85–93.

115 *The revelations stemming from:* H. G. Parker et al., "Genetic Struc-

ture of the Purebred Domestic Dog," *Science* 304, no. 5674 (2004): 1160–64.

116 *So vonHoldt and her colleagues:* The other strange thing about genetics is that it takes a heck of a lot of people to put together each scientific report. This paper has a total of thirty-six coauthors. B. M. vonHoldt et al., "Genome-wide SNP and Haplotype Analyses Reveal a Rich History Underlying Dog Domestication," *Nature* 464, no. 7290 (2010): 898–902.

"gene responsible for Williams-Beuren": Ibid.

117 *"Where Everybody Wants to Be":* ABC News online, *20/20,* https://abcnews.go.com/2020/video/williams-syndrome-children-friend-health-disease-hospital-doctors-13817012, undated.

I am especially fond of: "Cat-Friend vs. Dog-Friend," https://www.youtube.com/watch?v=GbycvPwr1Wg Nov 21, 2012.

119 *needed to figure out: Williams Syndrome Association,* "What Is Williams Syndrome?" https://williams-syndrome.org/what-is-williams-syndrome, undated.

123 *for this gene:* B. M. vonHoldt et al, "Structural Variants in Genes Associated with Human Williams-Beuren Syndrome Underlie Stereotypical Hypersociability in Domestic Dogs," *Science Advances* 3 (2017): e1700398.

124 *This adds grist to the mill:* M. E. Persson et al., "Sociality Genes Are Associated with Human-Directed Social Behaviour in Golden and Labrador Retriever Dogs," *PeerJ* 6 (2018): e5889.

"If they had tails": N. Rogers, "Rare Human Syndrome May Explain Why Dogs Are So Friendly," *Inside Science,* July 19, 2017, https://www.insidescience.org/news/rare-human-syndrome-may-explain-why-dogs-are-so-friendly.

Chapter 5: Origins

129 *"reared a hound with the greyest":* Arrian, "On Hunting," circa AD 145, in *Xenophon and Arrian on Hunting: With Hounds,* trans. A. A. Phillips and M. M. Willcock (Warminster, UK: Liverpool University Press, 1999).

130 *"The dog which was":* G. A. Reisner, "The Dog Which Was Hon-

ored by the King of Upper and Lower Egypt," *Bulletin: Museum of Fine Arts, Boston* 34, no. 206 (December 1936): 96–99, https://www.jstor.org/journal/bullmusefine.

132 *somewhat controversial:* L. Janssens et al., "A New Look at an Old Dog: Bonn-Oberkassel Reconsidered," *Journal of Archaeological Science* 92 (2018): 126–38.

133 *the eighteenth-century French naturalist:* Jean Léopold Nicolas Frédéric, Baron Cuvier, *Le Règne animal distribué d'après son organization.* Déterville libraire, 4 volumes (Paris: Imprimerie de A. Belin, 1817).

135 *The theme: how hunters:* There is a video of Moshe's interaction with his wolves available at this webpage: http://www.afikimproductions.com/Site/pages/en_inPage.asp?catID=10.

137 *Together with his wife:* Raymond Coppinger and Lorna Coppinger, *Dogs: A New Understanding of Canine Origin, Behavior, and Evolution* (Chicago: University of Chicago Press, 2002).

140 *The journalist Mark Derr:* Mark Derr, *How the Dog Became the Dog: From Wolves to Our Best Friends* (New York: The Overlook Press, 2013).

"the wolf [would have] voluntarily": Ibid., 131. The ugly truth is, free-roaming dogs in Zimbabwe get fully a quarter of their diet from human feces. J.R.A. Butler and J. T. du Toit, "Diet of Free-Ranging Domestic Dogs (Canis familiaris) in Rural Zimbabwe: Implications for Wild Scavengers on the Periphery of Wildlife Reserves," *Animal Conservation* 5, no. 1 (2002): 29–37.

143 *end of the last ice age:* Angela Perri, "Hunting Dogs as Environmental Adaptations in Jōmon Japan," *Antiquity* 90, no. 353 (October 2016): 1166–80. Angela Perri, *Global Hunting Adaptations to Early Holocene Temperate Forests: Intentional Dog Burials as Evidence of Hunting Strategies,* PhD Thesis, Durham University, 2013.

146 *"Sulu":* Strangely similar to the Latin word for wolf, lupu.

147 *but rather cry out:* The number one magazine for hunters who hunt with dogs in the United States is called *Full Cry,* in recognition of the central role that the dog's cry plays in the human-dog hunting team's success. Covers routinely feature a dog at the base of a tree, crying at something that has run up into the higher branches.

150 *resin copy of the bones:* The original bones are in a tiny museum at Kibbutz Ma'ayan Baruch in the far north of Israel.

152 *One archeologist proposed:* L. Larsson, "Mortuary Practices and Dog Graves in Mesolithic Societies of Southern Scandinavia," *Anthropologie* 98, no. 4 (1994): 562–75.

154 *With Stalin's death:* L. A. Dugatkin and L. Trut, *How to Tame a Fox (and Build a Dog): Visionary Scientists and a Siberian Tale of Jump-Started Evolution* (Chicago: University of Chicago Press, 2017).

"red in tooth and claw": A phrase often associated with Darwinian evolution, it actually originated a decade before Darwin with the poet Alfred, Lord Tennyson, in canto 56 of his poem *In Memoriam* (London: Edward Moxon, 1850).

155 *Ember, a member:* Dugatkin and Trut, *How to Tame a Fox,* 50–52.

Chapter 6: How Dogs Fall in Love

162 *The ethical implications:* D. E. Duncan, "Inside the Very Big, Very Controversial Business of Dog Cloning," *Vanity Fair,* September 2018.

163 *"Each puppy is unique":* B. Streisand, "Barbra Streisand Explains: Why I Cloned My Dog," *New York Times,* March 2, 2018.

165 *It was very hard:* Author interviews with Rich Hazelwood, August 15 and 16, 2018, Phoenix, Arizona.

166 *Little penguins are not:* Some authorities consider the Australian and New Zealand little penguins to be distinct species. In which case the NZ penguin is *Eudyptula minor* and the Australian species is *Eudyptula novaehollandiae.*

167 *"It got squashed":* Austin Ramzy, "Australia Deploys Sheepdogs to Save a Penguin Colony," *New York Times,* November 4, 2015, https://www.nytimes.com/2015/11/05/world/australia/australia-penguins-sheepdogs-foxes-swampy-marsh-farmer-middle-island.html.

"chic but discreet": Lisa Gerard-Sharp, *"Europe's Hidden Coasts: The Maremma, Italy," The Guardian,* May 22, 2017, https://www.theguardian.com/travel/2017/may/22/maremma-tucanny-coast-beaches-italy.

168 *Homer, in the* Odyssey: Book 14, http://classics.mit.edu/Homer/odyssey.14.xiv.html.

The first sheepdog trials: Barbara Cooper, "History of Sheepdog Trials," in *The Working Kelpie Council of Australia,* http://www.wkc.org.au/Historical-Trials/History-of-Sheepdog-Trials.php.

Unfortunately, the concept: Charles Darwin, *Voyage of the Beagle,* 2nd ed. (London: Murray, 1845), 75.

169 *Though Oddball became the star: Oddball,* directed by Stuart McDonald (Momentum Pictures, 2015).

Today the little penguins: Debbie Lustig, "Maremma Sheepdogs Keep Watch over Little Penguins," *Bark: The Dog Culture Magazine* 65 (July 2011), https://thebark.com/content/maremma-sheepdogs-keep-watch-over-little-penguins. Warrnambool City Council, "Maremma Dogs," 2018, http://www.warrnamboolpenguins.com.au/maremma-dogs. Maremma Sheepdog Club of America, "Maremma Sheepdog Breed History" 2014–2017, http://www.maremmaclub.com/history.html. Author interview with David Williams, August 9, 2018.

170 *"the method of education":* Darwin, *Voyage of the Beagle,* 150.

172 *The founder of the park:* "Eckhard H. Hess Dead at 69; Behavioral Scientist Authority," *New York Times,* February 26, 1986, https://www.nytimes.com/1986/02/26/obituaries/eckhard-h-hess-dead-at-69-behavioral-science-authority.html. Eckhard Hess, *Imprinting* (New York: Van Nostrand Reinhold, 1973).

175 *"like little wild animals":* D. G. Freedman, J. A. King, and O. Elliot, "Critical Period in the Social Development of Dogs," *Science* 133, no. 3457 (1961): 1016–17. John Paul Scott and John L. Fuller, *Genetics and the Social Behavior of the Dog* (Chicago: University of Chicago Press, 1965), 105. The two accounts of the experiment differ in how much human contact the dogs got. The *Science* paper claims ninety minutes per day. The book says only ten minutes per day. I presume the paper is more accurate, and the book was put together later, from memory.

180 *The person who now had:* M. Gacsi et al., "Attachment Behavior of Adult Dogs (Canis familiaris) Living at Rescue Centers: Forming New Bonds," *Journal of Comparative Psychology* 115, no. 4 (2001): 423–31.

182 *A willingness to explore:* E. N. Feuerbacher and C.D.L. Wynne, "Dogs Don't Always Prefer Their Owners and Can Quickly Form Strong Preferences for Certain Strangers over Others," *Journal of the Experimental Analysis of Behavior* 108, no. 3 (2017): 305–17.

183 *For example, Pete:* Sam Haysom, "This Story of a Heroic Dog Who Died Protecting His Owner Will Break Your Heart," *Mashable,* February 13, 2018, https://mashable.com/2018/02/13/dog-dies-after-protecting-owner-from-black-bear/#o4leySe3ekq0.

In 2016, Precious: "Service Dog Killed Trying to Protect Owner from Alligator in Florida," *CBS News,* June 24, 2016, https://www.cbsnews.com/news/service-dog-killed-trying-to-protect-owner-from-alligator-in-florida.

Jace Decosse's dog: Nadia Moharib, "'Hero' Dog Killed Defending Calgary Owner During Violent Home Invasion," *Edmonton Sun,* April 10, 2013, https://edmontonsun.com/2013/04/10/hero-dog-killed-defending-calgary-owner-during-violent-home-invasion/wcm/14a76ff4-9e1e-4ad8-9bd8-fb91a2245385.

I'm skeptical, however: Eric Knight, *Lassie Come-Home* (New York: Grosset & Dunlap, 1940).

184 *"I had a dog":* Charles Darwin, *The Descent of Man, and Selection in Relation to Sex,* vol. 1, 1st ed. (London: John Murray, 1871), 45.

186 *"genetics does not put":* John Paul Scott, "Investigative Behavior: Toward a Science of Sociality," in *Studying Animal Behavior: Autobiographies of the Founders,* ed. D. A. Dewsbury, 389–429 (Chicago: University of Chicago Press, 1985), 416.

7. Dogs Deserve Better

193 "We no longer recommend": Emphasis in the original. The Monks of New Skete, *How to Be Your Dog's Best Friend: The Classic Training Manual for Dog Owners* (Boston: Little, Brown, 2002).

We see animals strong-armed: I'm not saying that slip leads don't have legitimate purposes.

196 *Even some social animals:* C. Packer, A. E. Pusey, and L. E. Eberly, "Egalitarianism in Female African Lions," *Science* 293, no. 5530 (2001): 690–93.

197 *The crucial thing to realize:* L. D. Mech, "Alpha Status, Dominance, and Division of Labor in Wolf Packs," *Canadian Journal of Zoology* 77, no. 8 (November 1, 1999): 1196–203.

198 *But when a dominant dog:* F. Range, C. Ritter, and Z. Virányi,

"Testing the Myth: Tolerant Dogs and Aggressive Wolves," *Proceedings of the Royal Society: B. Biological Sciences,* 282 (2015): 20150220.

202 *"save your hugs":* S. Coren, "The Data Says 'Don't Hug the Dog!'" *Psychology Today: Canine Corner,* 2016, https://www.psychologytoday.com/blog/canine-corner/201604/the-data-says-dont-hug-the-dog.

203 *In Sweden, the law demands:* Svenska Kennelklubben, "Dog Owners in the City: Information About Keeping a Dog in Urban Areas," *Svenska Kennelklubben,* 2013, https://www.skk.se/globalassets/dokument/att-aga-hund/kampanjer/skall-inte-pa-hunden-2013/dog-owners-in-the-city_hi20.pdf.

204 *They have become the most:* D. van Rooy et al., "Risk Factors of Separation-Related Behaviours in Australian Retrievers," *Applied Animal Behaviour Science* 209 (December 1, 2018): 71–77. C. V. Spain, J. M. Scarlett, and K. A. Houpt, "Long-Term Risks and Benefits of Early-Age Gonadectomy in Dogs," *Journal of the American Veterinary Medical Association* 224, no. 3 (February 2004): 380–87.

205 *But that still leaves about a million:* The imprecision of these numbers is itself an issue. Nobody in the United States keeps a record of even how many shelters there are, never mind how many animals are in these facilities. Consequently, estimates have very wide margins of error. An excellent open-access article on all these issues is A. Rowan and T. Kartal, "Dog Population and Dog Sheltering Trends in the United States of America," *Animals: An Open Access Journal* 8, no. 5 (2018): 1–20.

208 *"Given that, in Italy":* S. Cafazzo et al., "Behavioural and Physiological Indicators of Shelter Dogs' Welfare: Reflections on the No-Kill Policy on Free-Ranging Dogs in Italy Revisited on the Basis of 15 Years of Implementation," *Physiology & Behavior* 133 (June 4, 2014): 223–29.

But I have seen nightmarish: P. D. Scheifele et al., "Effect of Kennel Noise on Hearing in Dogs," *American Journal of Veterinary Research* 73, no. 4 (2012): 482–89.

209 *When she was working on her PhD:* Sasha's formal first name is Alexandra.

210 *The dogs with the best chances:* A. Protopopova et al., "In-Kennel Behavior Predicts Length of Stay in Shelter Dogs," *PLOS ONE* 9, no. 12 (December 31, 2014): e114319.

213 *The fact that the fabled:* Anyone interested in what Pavlov really did, and why, needs to get hold of Daniel Todes's biography: D. P. Todes, *Ivan Pavlov: A Russian Life in Science* (Oxford, UK: Oxford University Press, 2014).

214 *In research that Lisa and I:* L. M. Gunter, R. T. Barber, and C.D.L. Wynne, "A Canine Identity Crisis: Genetic Breed Heritage Testing of Shelter Dogs," *PLOS ONE* 13, no. 8 (August 23, 2018): e0202633.

215 *As Bronwen Dickey explains:* B. Dickey, *Pit Bull: The Battle over an American Icon* (New York: Vintage, 2017).

217 *There were no losers:* L. M. Gunter, R. T. Barber, and C.D.L. Wynne, "What's in a Name? Effect of Breed Perceptions and Labeling on Attractiveness, Adoptions, and Length of Stay for Pit-Bull-Type Dogs," *PLOS ONE* 11, no. 3 (March 23, 2016): e0146857.

220 *Even the saluki:* H. G. Parker et al., "Genomic Analyses Reveal the Influence of Geographic Origin, Migration, and Hybridization on Modern Dog Breed Development," *Cell Reports* 19, no. 4 (2017): 697–708. B. M. vonHoldt et al., "Genome-wide SNP and Haplotype Analyses Reveal a Rich History Underlying Dog Domestication," *Nature* 464, no. 7290 (2010): 898–902.

Earlier than that: D. J. Brewer, T. Clark, and A. Phillips, *Dogs in Antiquity: Anubis to Cerberus — The Origins of the Domestic Dog* (Warminster, UK: Aris & Phillips, 2001).

221 *This program, which went on: Pedigree Dogs Exposed,* directed by Jemima Harrison, BBC TV, August 2008.

222 *"If the dog breeders insist":* Beverley Cuddy, "Controversy over BBC's Purebred Dog Breeding Documentary: BBC's Pedigree Dogs Exposed Strikes a Chord," *The Bark* 56 (September 2009), https://thebark.com/content/controversy-over-bbcs-purebred-dog-breeding-documentary.

The BBC documentary motivated: Patrick Bateson, *Independent Inquiry into Dog Breeding* (Cambridge, UK, 2010), https://www.ourdogs.co.uk/special/final-dog-inquiry-120110.pdf.

Britain's ten-thousand-plus: F.C.F. Calboli et al., "Population Structure and Inbreeding from Pedigree Analysis of Purebred Dogs," *Genetics* 179, no. 1 (May 1, 2008): 593–601.

223 *So far as the genetics:* Denise Powell, "Overcoming 20th-Century

Attitude About Cross Breeding," *Low Uric Acid Dalmatians World* (2016), https://luadalmatians-world.com/enus/dalmatian-articles/crossbreeding. L. L. Farrell et al., "The Challenges of Pedigree Dog Health: Approaches to Combating Inherited Disease," *Canine Genetics and Epidemiology* 2, no. 3 (February 11, 2015).

224 *Nobody would ever be able:* Valerie Elliott, "Fiona the Mongrel and a Spot of Bother at Crufts: 'Impure' Dalmatian Angers Traditionalists at the Elite Pedigree Dog Show," *Daily Mail,* March 6, 2011, https://www.dailymail.co.uk/news/article-1363354/Fiona-mongrel-spot-bother-Crufts-Impure-dalmatian-angers-traditionalists-elite-pedigree-dog-show.html.

If you are based in: US Government, *Animal Welfare Act* (Washington, DC: US Government Publishing Office, 2015), https://www.nal.usda.gov/awic/animal-welfare-act.

225 *For instance, the journalist Rory Kress:* Rory Kress, *The Doggie in the Window: How One Dog Led Me from the Pet Store to the Factory Farm to Uncover the Truth of Where Puppies Really Come From* (Naperville, IL: Sourcebooks, 2018).

Conclusion

231 *The vast majority of dogs:* Insurance Information Institute, "Dog-Bite Claims Nationwide Increased 2.2 Percent; California, Florida, and Pennsylvania Lead Nation in Number of Claims" (New York: Insurance Information Institute, 2018), https://www.iii.org/press-release/dog-bite-claims-nationwide-increased-22-percent-california-florida-and-pennsylvania-lead-nation-in-number-of-claims-040518.

Human-on-human violence: It should be acknowledged that this is an estimate of total costs, not just insurance payouts. H. R. Waters et al., "The Costs of Interpersonal Violence — an International Review," *Health Policy* 73, no. 3 (September 8, 2005): 303–15.

forty deaths each year: WISQARS Leading Causes of Death Reports, 1981–2017, National Center for Injury Prevention and Control, Centers for Disease Control, 2019, https://webappa.cdc.gov/sasweb/ncipc/leadcause.html.

INDEX

Page numbers in italics refer to text graphics.